DESERTIFICATION OF ARID LANDS

ADVANCES IN DESERT AND ARID LAND TECHNOLOGY AND DEVELOPMENT

Edited by: Adli Bishay
The American University in Cairo

William G. McGinnies
The University of Arizona

Editorial Advisory Board:

Harold Dregne
Texas Tech University
(U.S.A.)

M. Evanari
Hebrew University of Jerusalem (Israel)

Henri N. Le Houreou
Montpellier-Cedex
(France)

Enrique Campos-Lopez
Centro de Investigacion en Quimica Aplicada
(Mexico)

H. S. Mann
Ludhiana, Punjab
(India)

Nina T. Nechaeva
Desert Institute
USSR Academy of Sciences
Ashkabad (USSR)

Ray Perry
Institute of Earth Resources
(Australia)

Abdu Shatta
Desert Institute
El Mataria (Egypt)

Volume 1   APPLICATIONS OF SCIENCE AND TECHNOLOGY FOR DESERT DEVELOPMENT
                Edited by Adli Bishay and William G. McGinnies

Volume 2   REALISTIC PLANNING FOR ARID LANDS
                Natural Resource Limitations to Agricultural Development, by W. Gerald Matlock

Volume 3   DESERTIFICATION OF ARID LANDS
                by H. E. Dregne

Volume 4   IMPROVEMENT OF DESERT RANGES IN SOVIET CENTRAL ASIA
                Edited by Nina T. Nechaeva

Other volumes in preparation

ISSN: 0142-5889

This book is part of a series. The publisher will accept continuation orders, which may be cancelled at any time and which provide for the automatic billing and shipping of each title in the series upon publication. Please write for details.

DESERTIFICATION OF ARID LANDS

H. E. Dregne
Texas Tech University

GB
611
.D73
1983
West

harwood academic publishers
chur · london · paris · new york

Published by Harwood Academic Publishers GmbH. All Rights Reserved.

Harwood Academic Publishers

Poststrasse 22
7000 Chur
Switzerland

P.O. Box 197
London WC2E 9PX
England

58, rue Lhomond
75005 Paris
France

P.O. Box 786
Cooper Station
New York, NY 10276
United States of America

Published 1983
Second Printing 1985

Copyright © 1983 Library of Congress

Library of Congress Cataloging in Publication Data

Dregne, H. E. (Harold E.)
  Desertification of arid lands.

  (Advances in desert and arid land technology and development, ISSN 0142-5889; v.3)
  Includes bibliographical references and index.
  1. Desertification. I. Title. II. Series.
GB611.D73 1983   333.73'13   83-8396
ISBN 3-7186-0168-0

ISBN 3-7186-0168-0. ISSN 0142-5889. No part of this book may be reproduced or utilized in any form or by any means, electronic or mechanical, including photocopying and recording, or by any information storage or retrieval system, without permission in writing from the publishers. Printed in the United States of America.

TABLE OF CONTENTS

Preface to the Series		vii
Preface		ix
1.	Introduction	1
	International Directions	1
	Definition	4
	Desert Spread and Droughts	7
	Historical Development	8
	Extent of Desertification	15
	Impact of Desertification	19
	References	25
2.	Desertification Processes	28
	Principal Processes	29
	Desertification and Land Use	36
	Grazing Land	36
	Dryland Cropping	41
	Irrigated Agriculture	46
	Mining	50
	Tourism and Recreation	51
	Woodcutting	52
	Urban Development	53
	Conclusions	54
	References	55
3.	Desertification Indicators	58
	Grazing Lands	59
	Rainfed Croplands	63
	Irrigated Lands	65
	Desertification Indicators	66
	Critical Indicators	68
	Conclusions	93
	References	94

4.	Cause and Effect	96
	Climate Change	96
	Rangelands	97
	Dryland Farming	106
	Irrigation	110
	Mining	119
	Recreation	122
	Communities	125
	Conclusions	127
	References	131
5.	Prevention and Reversal of Desertification	136
	Combatting Desertification	137
	Management Practices	143
	Desertification Control	160
	Conclusions	165
	References	167
6.	Occurrence of Desertification	171
	Africa	174
	Asia	185
	Australia	192
	North America	199
	South America	209
	Spain	215
	Conclusions	220
	References	222
Index		227

Preface to the Series

The general purpose of the Advances in Desert and Arid Land Technology and Development series is to bring an interdisciplinary approach to the problems of desert technology and development. The series will include original work and review articles covering science, technology, engineering, agriculture, architecture, sociology, management and economics of desert and arid land utilization and development. Both invited and unsolicited papers of high merit will be selected for inclusion in the series.

W. G. McGinnies	Adli Bishay
Tucson	Cairo

PREFACE

Desertification is a term that became widely known following the environmental destruction and human suffering caused by the 1969-1973 drought in the sub-Saharan Sahel. Curiously enough, in view of the drought-inspired motivation for the 1977 United Nations Conference on Desertification, the consultants who prepared the background documents concluded that drought was not the cause of desertification, only a contributor to it; rather, unwise human activity was identified as the culprit.

The distinction between attributing desertification to drought or to human activity is an important one. If drought is the cause, we can say that there is little man can do to counter a natural phenomenon, leaving us with an excuse to do nothing. If, however, man is the cause, then man has the opportunity to undo the damage he has done or at least to prevent further deterioration. Few knowledgeable persons now contend that the present-day climate differs significantly from the climate of a thousand or more years ago; changes have occurred in the length and amplitude of wet and dry cycles but the long-time average remains essentially constant. Worse droughts than that of 1969-73 have occurred in the past and worse droughts will come again, sooner or later. Their impact will depend upon the steps we have taken to protect the land resource.

Desertification is not a recent development. Plato

is quoted by Arnold J. Toynbee as having bemoaned nearly 2000 years ago the fact that Grecian Attica was a "mere relic of the original country. . .. All of the rich soft soil has moulted away, leaving a country of skin and bones. . .". Tree cutting, overgrazing, and water erosion were the desertification factors that were responsible for the deplorable condition in Attica. The same factors were operative in other countries around the Mediterranean Sea, with similar results. Plato's comment about rains that formerly soaked into the soil being lost as runoff in his day describes the onset of edaphic aridity that gives the false impression of a climatic change.

For the most part land degradation and desertification are synonyms. Perhaps the principal merit of referring to land degradation as desertification is the mental image that the term conjures: the end point of land degradation is a man-made wasteland. Unfortunately, too many people think desertification only refers to the extreme condition of wasteland formation, a condition that affects only a small part of the earth's surface, thus far.

Among the many enjoyable consequences of my studies on desertification has been the opportunity to learn from my colleagues. I am especially grateful for the insights I gained from Manuel Anaya Garduño, Gaafar Karrar, Mohamed Kassas, Victor Kovda, Fernando Medellin Leal, Jack Mabbutt, Monique Mainguet, Ray Perry, Boris Rozanov, and Gilbert White.

H. E. Dregne

1. INTRODUCTION

Desertification of the arid lands of the world has been proceeding - sometimes rapidly, sometimes slowly - for more than a thousand years. It has caused untold misery among those most directly affected, yet environmental destruction continues. Until recently, few if any lessons seemed to have been learned from the past, in part because the problem was an insidious one that went unrecognized in its early stages or was seen as a local one affecting only a small population, in part because new land was always available to start over again. As long as remedial action could be deferred by moving on to new frontiers, land conservation had little appeal. It was not until the 20th century - when easy land expansion came to an end - that governments and people finally realized that continued careless degradation of natural resources threatened their future.

INTERNATIONAL DIRECTIONS

The decade of the 1950's witnessed the first worldwide effort to call attention to the problems and potentials of arid regions. It started when the United Nations Educational, Scientific, and Cultural Organization (UNESCO) launched its Major Project on Scientific Research on Arid Lands in 1951. That project led to publication of a newsletter, the provision of funds for establishing and strengthening arid land research institutes, organization of conferences and symposia, and publication of a series of research reviews and special reports on a wide range of topics. The Major Project was terminated in 1962 and the arid land program was merged with the broader UNESCO natural resource program.

The impetus generated by the UNESCO project led to expanded interest in, and support of, arid lands studies throughout the world. By 1970, knowledgeable scientists were fully aware of the magnitude of the land degradation that had taken place in the past and that was becoming even more serious as population pressures increased. Still, application of remedial measures to combat land degradation was slow, by and large, in the absence of any sense of crisis. The situation was similar, on a world basis, to the apathy that prevailed in the United States toward accelerated soil erosion before Hugh Hammond Bennett dramatized to political leaders the damage erosion had done. The great drought of the 1930's came along at exactly the right time to raise the specter of wind erosion ruining land and people's lives in the Great Plains. As a result of Bennett and the drought, the U.S. Congress established the Soil Conservation Service to assist farmers to combat soil erosion. A pall of West Texas dust hanging over Washington, D.C. in May, 1934 helped persuade congressmen that wind erosion was a problem deserving immediate attention.

In the early 1970's, another drought served to focus world attention on the land degradation problem called desertification: the 1969 to 1973 drought in the African Sahel. Recognition of the severity of the drought's affect on six countries on the southern border

INTRODUCTION 3

of the Sahara (Mauritania, Senegal, Mali, Upper Volta, Niger, and Chad) was slow to develop. Droughts, after all, were not unusual in the Sahelian countries; an equally bad or even worse one had struck the same region during the years from 1911 to 1914, and several other droughts had occurred before and after that time. What made this one different was the severe overgrazing and vegetation destruction that had accompanied increases in livestock numbers and expansion of cultivated land into marginal areas in the years before 1969. When drought struck, weakened fodder plants and bare soil were in no condition to cope with the climatic shock.

Among the after-effects of the human toll and the millions of livestock that died due to the drought was the call by the United Nations General Assembly for the convening of an international conference on desertification. The United Nations Conference on Desertification was held in Nairobi, Kenya in August and September of 1977, attended by representatives of nearly 100 countries and many international organizations, governmental and nongovernmental. In addition to providing a forum by which the world's attention was drawn to the problem of desertification, the Conference addressed the problem of combatting the land destruction that had occurred in the arid regions. A Plan of Action was developed and approved by the delegates (United Nations, 1978). Responsibility for following up and coordinating the implementation of the Plan of Action to Combat Desertification was entrusted to the United Nations Environment Program.

Ironically, association of the Sahelian drought with

desertification has led to the common misconception that drought causes desertification. In truth, droughts do not cause desertification, they only exaggerate the harmful effects of improper land management.

DEFINITION

Desertification is a term that has been in use since at least 1949 when Aubréville, a perceptive and well informed botanist and ecologist, published a book on "Climats, Forêts, et Désertification de l'Afrique Tropicale" (Aubréville, 1949). Aubréville thought of desertification as the changing of productive land into a wasteland as the result of ruination by man-induced soil erosion. He associated it with the humid and sub-humid tropics where he worked. The causes of land destruction were tree cutting, indiscriminate use of fire, and cultivation, which exposed the soil to water and wind erosion. Desertification was not the result of the Sahara spreading outward but of localized activity that could begin anywhere. Aubreville was quite clear in his conclusion that desertification in tropical Africa was due to man's activity, and that there had been no significant climatic change during the past thousand or more years. Most of that destructive activity has occurred within recent historic time by the action of agricultural populations.

Despite the fact that a world conference has been held on the subject, there is no generally accepted definition of desertification. Among different people, the term may refer variously to 1) degradation of grazing lands, 2) destruction of vegetative cover, 3) wind erosion and moving sand dunes, 4) turning productive land into a wasteland by whatever means, and 5) degradation of vegetative and soil resources. The latter interpretation is the one used by consultants to the United Nations Conference on Desertification (United Nations, 1977) and is the one used here. Additionally, desertification is generally considered to be a phenomenon confined to the dry lands of the world.

The definition used here is the following:

> Desertification is the impoverishment of terrestrial ecosystems under the impact of man. It is the process of deterioration in these ecosystems that can be measured by reduced productivity of desirable plants, undesirable alterations in the biomass and the diversity of the micro and macro fauna and flora, accelerated soil deterioration, and increased hazards for human occupancy.

This definition says that desertification is a land degradation process which may involve a continuum of change, from slight to very severe desertification of the plant and soil resource, and is due to man's activities. Nothing in the definition restricts desertification to the arid regions. However, in accordance with the general view, this monograph is concerned only with descrtification of arid lands. Figure 1.1 delimits the arid regions. The map is re-drawn from a UNESCO map (UNESCO, 1977).

Desertified land is not necessarily useless waste land, although many people associate desertification with barren forbidding landscapes resembling Death Valley or

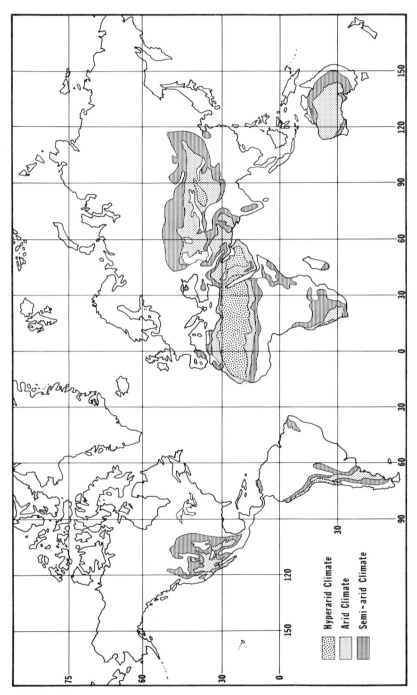

Figure 1.1. Distribution of arid region climates

the Sahara. All stages of desertification, from slight to very severe can be seen in most arid region countries. Slightly desertified land can usually be restored to full productivity rather easily; very severely desertified land may be so badly degraded that restoration is not economically feasible. The principal desertification processes are degradation of vegetative cover, water and wind erosion, soil compaction, and salinization and waterlogging. The results of desertification are a lower carrying capacity for animals and people, a reduction in crop production potential, an increase in environmental deterioration, increased flooding of lower lying land, and reduced capacity to support human life.

DESERT SPREAD AND DROUGHTS

A common misapprehension about desertification is that it spreads from a desert core, like a ripple on a pond. The truth is that land degradation can and does occur far from any climatic desert; the presence or absence of a nearby desert has no direct relation to desertification. Desertification usually begins as a patch on the landscape where land abuse has become excessive. From that patch, which might be around a watering point or in a cultivated field, land degradation spreads outward if the abuse continues. Ultimately the patches may merge into a homogeneous area, but that is unusual on a large scale.

A second misconception is that droughts are responsible for desertification. Droughts do increase the likelihood that the rate of degradation will increase on non-irrigated land if the carrying capacity is exceeded. However, well managed land will recover from droughts with minimal adverse effects when the rains return. The deadly combination is land abuse during good periods and its continuation during periods of deficient rainfall. Under those circumstances, weakened plants and exposed soils may be unable to withstand the onslaught of drought when combined with overgrazing, cultivation, and wood cutting. The famous satellite photograph taken during the 1970's Sahel drought of a green rectangle (the Ekrafane Ranch) surrounded by brown denuded land in eastern Niger demonstrated dramatically the value of good range management during favorable and unfavorable years.

HISTORICAL DEVELOPMENT

Land degradation is by no means a new problem, despite the attention focussed upon it in recent years. In some quarters, there is a tendency to blame desertification upon land pressures generated by the population explosion of the middle 20th century. While the rapidly expanding population has greatly exacerbated the situation, desertification is not a new phenomenon.

Historical evidence shows that serious and extensive land deterioration occurred several centuries ago in the arid zones, principally in three regions: around the Mediterranean Sea, the Mesopotamian Valley, and the loessial plateau of China. There were other places where destructive changes in soil and plant cover had occurred but they either were small in extent or are not well known.

INTRODUCTION

Mediterranean Region

Accounts of Roman water conservation and irrigation techniques in North Africa and the eastern Mediterranean indicate that they had developed a high degree of proficiency in raising crops under conditions of limited rainfall. Ruins of cities now surrounded by desolate waste land in Libya and Tunisia attest to the changes that have occurred in a little more than one thousand years. Irrigation and terrace systems that once must have been engineering marvels fell into disuse when the empire weakened and invading hosts pillaged the land. The Pax Romana made possible a fruitful agriculture; when peace ended and people withdrew to protected strongholds, some of the untended land simply reverted to what it had been originally. Much of the crop land, however, eroded away under the forces of water and wind, leaving gullies and sand dunes to mark where it had been (Bennett, 1939). Heichelheim (1956) believed that decay of the Roman agricultural systems occurred some time after the fifth and sixth centuries A.D. It is an unfortunate fact that water erosion on terraced land frequently is worse after the terraces fall into disrepair than when there were not terraces at all. Terraces serve to concentrate water, which is good when they are maintained but damaging when they are not.

All around the periphery of the Mediterranean Sea, beginning perhaps with the Phoenicians in the east, mountain slopes were deforested to provide timber for ships, temples, and ore smelters and to clear land for cultivation (Beals, 1965; Bennett, 1939; Mikesell, 1960; Naveh and Dan, 1973; Tomaselli, 1977). Destruction of the stately cedars of Lebanon is the classical example of how tree cutting leads to land degradation (Mikesell, 1969). With the loss of the protective cover, soil erosion was accelerated and landscapes altered, giving the false impression that the climate had become more arid.

As long ago as 2600 B.C., Egyptians were known to have depended upon the Phoenician forests to supply them with timber for ships and temples and for the resins used in mummification. Demand for Levantine cedars, firs, and pines was so great that by the time of the Mameluke regime (1250-1517 A.D.) timber was regarded as a scarce strategic material and its cutting was placed under the nominal control of the sultan (Mikesell, 1969). A brief recovery occurred in the 18th century when the area was depopulated as the result of social upheaval but the final blow was struck at the beginning of the 20th century when virtually the last remnants were cut for use on railroads for fuel and roadbed construction. Today only a few sacred groves remain to show how things once were. Even though the climate is the same as it was when the cedars flourished, the possibility of reforestation is slight as long as people must rely on wood for fuel and goats are free to eat every green plant that manages to germinate (Mikesell, 1969).

Along with destruction of forests went the expansion of cultivated land up the mountain slopes. Early settlers apparently were able to hold the soil in place by constructing terraces. In later years, as armies moved back and forth across the countryside and as community fortunes rose and fell, maintenance of terraces and other

water control structures often was neglected. The
inevitable result was accelerated soil erosion, leading
to the barren landscapes that now mark the Levant
(Stallings, 1957). Occasionally, well maintained terraces
demonstrate that the land is capable of being cropped
for thousands of years if properly cared for.

Tomaselli (1977) details the degradation of timber
land in the Mediterranean countries, spreading from east
to west as new colonies were established. He points out
that the Mediterranean maquis type of vegetation,
consisting of non-climax woody species (scrub oak,
heath, etc.) now is the common vegetative cover on what
once was forested land. Maquis vegetation is so common
that it frequently is believed to be the climax vegetation
everywhere. In fact, around the Mediterranean it is
sometimes difficult to determine what the original
vegetation was because man drastically altered the plant
and soil cover centuries ago.

Mesopotamia

The most significant instance of desertification
resulting from waterlogging and salinization in ancient
times is found in the lower valleys of the Tigris and
Euphrates rivers. Present-day Iraq is still trying to
recover from damage done more than 1,500 years ago in
the region known as Mesopotamia.

Jacobsen and Adams (1958) have chronicled the
history of Mesopotamian irrigation and its effect on crop
production. They noted that salinity problems became
serious for the first time in about 2400 B.C. in the

Girsu area of southern Iraq. That phase lasted until at
least 1700 B.C. It was followed by a less serious period
of salinization in central Iraq three thousand years
ago. The last major period of salt damage began eight
hundred years ago east of Baghdad. Some of the long-
abandoned fields recovered when the water table dropped,
while others - particularly in the south - seem never to
have returned to their earlier productive condition.

In addition to salinity and water table problems,
irrigation systems in Mesopotamia were plagued with
siltation of canals. The heavy load of sediment coming
off eroded uplands made it mandatory that canals be
cleaned frequently. How much of the silt was due to
timber cutting, overgrazing, and cultivation on the watersheds of the rivers is not known. It probably was
significant. In any event, sedimentation apparently
required a temporarily successful major modification in
the system of water withdrawal from the 300 kilometer
Nahrwan canal around 700 A.D. Ultimately, siltation
proved to be too intractable to handle and much of the
land had to be abandoned.

Loessial Plateau of China

Loess is wind-deposited soil particles and aggregates
of silt and clay, the bulk of which has a diameter of
0.01 to 0.05 mm. Loessial soils are among the most
productive soils in the world. They are extensive in the
southern part of the Soviet Union, the central United
States, eastern Argentina, and northern China. The
thickest deposits are in semi-arid China, in and around

the big bend of the Yellow River. In some places the loess there is said to be more than 100 meters thick (Barbour, 1926).

Although loess is inherently productive, it is also highly susceptible to water erosion. The soil particles are of a size and shape that makes them adhere loosely to one another so that water erosion forms deep, vertically-sided gullies. That susceptibility to erosion is responsible for the occurrence in China of some of the most serious land degradation anywhere in the world. One estimate is that gullies make up 26,000 square kilometers of the total 600,000 square kilometers of loess land (Kuo, 1976). Gully and sheet erosion in the loessial region is responsible for most the tremendous load of silt carried by the Yellow River.

Thorp (1936) was of the opinion that soil erosion was serious on the loess plateau several centuries ago. Southern Shansi Province is the original home of the Chinese civilization, which means that the land has been cultivated for millenia. It now supports an ever increasing population of more than 50,000,000 people, and erosion continues. The magnitude of erosion on the terraced land is hard to believe. In 1934, Thorp observed one field on a hillside in Shansi Province where the entire 10 centimeter plow depth was swept away in a single short rainy period. Even with that extremely rapid erosion, it takes many centuries to produce the severely gullied landscape that now typifies the loessial plateau.

Last One Hundred Years

Desertification in Africa and elsewhere began long before the 1969-1973 drought struck the Sahel. Stebbing (1937a, 1937b) was the most persistent of those sounding the alarm over the rapid degradation of the Sahelian and Sudanian vegetative zones. He looked upon desertification in West Africa as vegetation degradation which leads to erosion and, as a last stage, to barren sand or rock. That process is hastened, he contended, by blown sand from the Sahara being deposited on the degraded land. It was this latter activity that led him to refer to the "encroaching Sahara", a term he later regretted using (Stebbing, 1938). He disclaimed any attempt to give the idea that the Sahara was a vast sand field advancing in great waves like the incoming tide of a sea. The latter concept, however, has proved so attractive to numerous writers on desertification that it now represents a common view on the subject (Cloudsley-Thompson, 1974). There apparently is something fascinating about the idea of an expanding desert threatening mankind.

Warnings similar to those of Stebbing for the Sahel were made by other scientists in southern Africa, North America, South America, Asia, and Australia during the 1920's and 1930's. Research has been undertaken in many countries to develop techniques of grazing management and soil and water conservation that would halt and reverse desertification. As a result, there is now a good understanding of the basic principles of land conservation. Field application of those principles has been slow, unfortunately, and land degradation continues to undermine

INTRODUCTION

efforts to improve human well-being.

EXTENT OF DESERTIFICATION

Figure 1.2 shows the extent of desertification in the arid lands of the world. Continental desertification maps can be found in Chapter 6. The classification system used in the preparation of the map is a simple qualitative one utilizing four levels of desertification: slight, moderate, severe, and very severe. Criteria used for placing land in one of the four desertification classes are presented in Table 1.1. Destruction of plant cover is the major desertification factor for grazing land, soil erosion is the major factor for rainfed cropland, and salinization and waterlogging for irrigated land. Grazing lands may exhibit both accelerated erosion and deterioration of the plant cover.

Based upon the delineations of Figure 1.2, the area of land in the various desertification classes is given in Table 1.2, by continents.

As Table 1.3 indicates, about 80 percent of the agricultural land in the arid regions of the world has experienced moderate or worse desertification. Irrigated lands are least affected and rangelands are most affected, although the latter are only slightly worse off than the rainfed croplands. Rangelands are primarily afflicted with overgrazing, woodcutting, soil compaction, and erosion, whereas rainfed croplands mainly have erosion and soil compaction problems. The dominant problems in irrigated land are waterlogging and salinization.

The total area of irrigated, rainfed, and grazing lands in Table 1.3 adds up to only about 41 million square

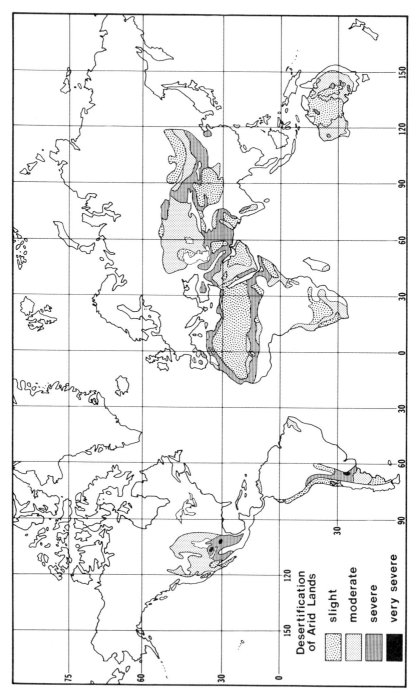

Figure 1.2. World status of desertification of arid lands

TABLE 1.1. Criteria for estimating degree of desertification.

Desertification class	Plant cover	Erosion	Salinization or waterlogging (irrigated land)	Crop yields
Slight	Excellent to good range condition class	None to slight	$ECe \times 10^3 < 4$ mmhos	Crop yields reduced less than 10 percent
Moderate	Fair range condition class	Moderate sheet erosion, shallow gullies, few hummocks	$ECe \times 10^3$ 4-8 mmhos	Crop yields reduced 10-50 percent
Severe	Poor range condition class	Severe sheet erosion, gullies common, occasional blow-out area	$ECe \times 10^3$ 8-15 mmhos	Crop yields reduced 50-90 percent
Very severe	Land essentially denuded of vegetation	Severely gullied, or numerous blow-out areas	Thick salt crust on nearly impermeable soils	Crop yields reduced more than 90 percent

TABLE 1.2. Land area in four desertification classes, by continent

Continent	Desertification class	Land area, sq. km.	Percent of arid lands
Africa	Slight	12,430,000	71.7
	Moderate	1,870,000	10.8
	Severe	3,030,000	17.5
	Total	17,330,000	100.0
Asia	Slight	7,980,000	50.9
	Moderate	4,480,000	28.6
	Severe	3,210,000	20.5
	Total	15,670,000	100.0
Australia	Slight	2,330,000	36.6
	Moderate	3,510,000	55.2
	Severe	520,000	8.2
	Total	6,360,000	100.0
North America	Slight	440,000	9.9
	Moderate	2,720,000	61.5
	Severe	1,200,000	27.1
	Very severe	67,000	1.5
	Total	4,427,000	100.0
South America	Slight	1,340,000	43.6
	Moderate	1,050,000	34.1
	Severe	680,000	22.1
	Very severe	6,000	0.2
	Total	3,076,000	100.0
Europe (Spain)	Moderate	140,000	70.0
	Severe	60,000	30.0
	Total	200,000	100.0
All continents	Slight	24,520,000	52.1
	Moderate	13,770,000	29.3
	Severe	8,700,000	18.5
	Very severe	73,000	0.1
	Total	47,063,000	100.0

INTRODUCTION   19

kilometers, 6 million less than the total area of arid
lands of the world (Table 1.2). The difference is the
area of hyperarid climates which are not considered to
be rangelands because their productivity is close to zero.
Most of that 6 million hectares is in the Sahara, the
Arabian peninsula, and the Takla Makan desert.

TABLE 1.3.   Extent of moderate to very severe desertification of agricultural land in arid regions

Land use	Total area ha	Area affected by desertification ha	Percent of total area
Irrigation	126,300,000	27,100,000	21
Rainfed Crops	224,400,000	173,100,000	77
Range	3,751,100,000	3,071,600,000	82
Total	4,101,800,000	3,271,800,000	80

IMPACT OF DESERTIFICATION

   Land degradation, in whatever form, is important only
insofar as it affects people. Of the estimated population
of 700,000,000 people residing in the arid regions, about
78,000,000 (Kates et al, 1977) live in areas where severe
desertification has occurred (Table 1.4). Out of that
78,000,000 people, about one-third may be in a position,
due to high income and other advantages, to avoid the
worst consequences of desertification. That leaves about
50,000,000 people who have already experienced a major
loss in their ability to support themselves. An unknown

TABLE 1.4. Estimates of populations -- According to livelihood -- In areas recently undergoing severe desertification* (In thousands)

Region	Total population	Urban based	Cropping based	Animal based
Mediterranean Basin	9,820	2,995 31%	5,900 60%	925 9%
Sub-Saharan Africa	16,165	3,072 19%	6,014 37%	7,079 44%
Asia and the Pacific	28,482	7,740 27%	14,311 54%	6,431 19%
Americas	24,079	7,683 32%	13,417 56%	2,979 12%
TOTAL	78,546	21,490 27%	39,642 51%	17,414 22%

*Source: Kates et al, 1977

percentage of that 50 million will have to abandon their agricultural way of life and join the overcrowded cities to seek relief.

The foregoing numbers refer only to people living in the severely desertified areas. At least eight times as many reside in the moderately desertified portions of the arid lands. If one-third of this group also can be expected to escape the impact of desertification, that leaves over 400,000,000 who are affected. Since the moderately desertified areas have suffered up to 50 percent loss of productivity from the land - measured in crop and animal production - it is apparent that land degradation has curtailed the livelihood of several hundred million people of the arid regions.

One conservative estimate of the amount of lost wheat production worldwide that could be attributed to desertification is 23,000,000 metric tons each year (Dregne, 1978). That much wheat is sufficient to meet the caloric requirements of 80,000,000 people for one year. For all of the arid regions, the current value of lost agricultural production is estimated to be about $26 billion each year (Table 1.5).

A seeming paradox that causes some questioning of the effect of desertification on crop yields can be observed in the wheat growing Great Plains region of the United States. There, wheat yields have remained fairly constant or actually increased over the past forty years, despite continued soil erosion and land degradation. The question arises of why yields are not falling if desertification is continuing. The explanation is that improvements in crop varieties, soil and water management

TABLE 1.5. Value of current annual loss of agricultural production due to past desertification

Type of land use	Annual loss*
Irrigated land	$ 6.6 billion
Rainfed cropland	12.3 billion
Rangeland	7.4 billion
TOTAL	$26.3 billion

*20% of irrigated land potential production, 30% of rainfed cropland, and 60% of rangeland.

practices, pest control, and fertilizer have, thus far, offset the effects of land deterioration. If those improvements through research do not continue into the future, crop yields can be expected to decline in the semi-arid United States and in other countries.

Every year, part of the cultivated land becomes so badly desertified that it no longer provides an income sufficient to offset costs. That land is unlikely to be abandoned. Instead, it continues to be used as before, even though the benefit/cost ratio is less than 1, or it is put to a less profitable use such as grazing. The amount of land which annually is reduced to the level of a zero or negative net return (profit) is estimated to be about 20,246,000 hectares (Table 1.6). No figures are available on the amount of land which is retired from any use (completely abandoned) but it must be very small. Rainfed cropland which is abandoned after a series of crop failures will still be used for grazing, in the great

INTRODUCTION 23

majority of instances. It is not "lost to the desert." Only man-induced moving sand dunes are liable to be completely unused. And they occupy little of the earth's land area.

TABLE 1.6. Area of land deteriorating annually to level of zero or negative net return

Type of land use	Annual deterioration ha
Irrigated land	546,000
Rainfed cropland	2,000,000
Rangeland	17,700,000
TOTAL	20,246,000

The cost of combatting desertification is given in Table 1.7 for the three major land uses. Total costs are greatest on rangelands because they are by far the most extensive. Costs per hectare, however, are highest for irrigated land and lowest for rangeland. Staggering though the figure of $141 billion is, it is only about five and one-half times the value of lost annual production (Table 1.5), which means that the costs could be recouped in less than ten years after the potential productivity had been achieved. However, not all desertified land is economically worth reclaiming. The benefit/cost ratio for combatting desertification is less than 1 on about 75 percent of the rangeland and 30 percent of the rainfed cropland (Table 1.8). Reclamation efforts, therefore, should be concentrated on irrigated land and the more favorable rainfed and range lands. The 30 percent of the rainfed cropland that does not give a positive net return (benefit/cost ratio greater than 1) is mostly the climatically marginal lands that were good grazing lands but are poor croplands. The poorer rangelands are in the drier regions or have suffered such severe sheet erosion or dune invasions that forage establishment and dune stabilization are difficult.

TABLE 1.7. Cost of combatting desertification on agricultural lands

Type of land use	Area desertified (ha)	Cost of improvement per hectare	total
Irrigated cropland	27,100,000	$750	$ 20.3 billion
Rainfed cropland	173,100,000	$250	43.2 billion
Rangeland	3,071,600,000	$ 25	77.5 billion
TOTAL			$141.0 billion

TABLE 1.8. Percent of land capable of giving positive net return from improvement

Type of land use	Percent of land
Irrigated land	100
Rainfed cropland	70
Rangeland	25

INTRODUCTION

Despite the fact that 75 percent of the rangelands are expected to not repay the cost of improving them, there may be good social reasons for investing in improvement programs. Among them are the desirability of slowing the rural exodus to the cities, equity in the expenditures of development funds, prevention of further deterioration of a natural resource, and providing employment opportunities during certain periods of the year. Reasons for improving the worst rainfed croplands are harder to justify. The climatically marginal croplands are unlikely to ever provide a satisfactory income for the people living on them even if further degradation is halted.

The high cost of coping with serious land damage, both in terms of lost income as well as in the expense of reclamation, makes prevention a much cheaper course of action. Unfortunately, that course is seldom followed.

REFERENCES

Aubréville, A. 1949. Climats, Forêts, et Désertification de l'Afrique Tropicale. Societe de Editions Geographiques, Martime et Coloniales. Paris. 255 p.

Barbour, George B. 1926. The loess of China. Smithsonian Institution, Annual Report of the Board of Regents, pp. 279-296.

Beals, E. W. 1965. The remnant cedar forests of Lebanon. Journal of Ecology, 53(3): 679-694.

Bennett, Hugh Hammond. 1939. Soil Conservation. McGraw-Hill Book Company, Inc., New York. 993 p.

Cloudsley-Thompson, J. L. 1974. The expanding Sahara. Environmental Conservation, 1: 5-13.

de Crespigny, R. R. C. 1971. China. St. Martin's Press, New York, N.Y. 235 p.

Dregne, H. E. 1978. The effect of desertification on crop production in semi-arid regions. In: Glen H. Cannell (editor), Proceedings of an International Symposium on Rainfed Agriculture in Semi-Arid Regions, University of California, Riverside, California, pp. 113-127.

Heichelheim, Fritz M. 1956. Effects of classical antiquity on the land. In: William L. Thomas, Jr. (editor), Man's Role in Changing the Face of the Earth. University of Chicago Press, Chicago, Illinois. pp. 165-182.

Jacobsen, Thorkild and Robert M. Adams. 1958. Salt and silt in ancient Mesopotamian agriculture. Science 128: 1251-1258.

Kates, Robert W., Douglas L. Johnson, and Kirsten Johnson Hering. 1977. Population Society and Desertification. United Nations Conference on Desertification. A/CONF. 74/8. 62 p.

Kuo, Leslie T. C. 1976. Agriculture in the People's Republic of China. Praeger Publishers, New York. 288 p.

Mikesell, Marvin W. 1960. Deforestation in northern Morocco. Science 132: 441-454.

Mikesell, Marvin W. 1969. The deforestation of Mount Lebanon. Geographical Review 59: 1-28.

Naveh, Zev and Joel Dan. 1973. The Human Degradation of Mediterranean Landscapes in Israel. In: F. di Castri and H. A. Mooney (editors), Mediterranean Type Ecosystems. Springer-Verlag, Berlin, pp. 373-390.

Stallings, J. H. 1957. Soil Conservation. Prentice-Hall, Inc., Englewood Cliffs, N.J. 575 p.

Stebbing, E. P. 1937a. The encroaching Sahara: The threat to the West African colonies. The Geographical Journal, 86: 506-519.

Stebbing, E. P. 1937b. The threat of the Sahara. Journal of the Royal African Society. Extra supplement to Vol. 36. 35 p. 25 May 1937.

Stebbing, E. P. 1938. The advance of the desert. The Geographical Journal 91: 356-359.

Thorp, J. 1936. Geography of the Soils of China. The National Geological Survey of China. Nanking, China.

Tomaselli, Ruggero. 1977. The degradation of the Mediterranean Maquis. Ambio, 6: 356-362.

UNESCO. 1977. World Distribution of Arid Regions. Map scale: 1/25,000,000. UNESCO, Paris.

United Nations. 1977. Desertification: An Overview. United Nations Conference on Desertification. A/CONF. 74/1, Nairobi, Kenya. 79 p.

United Nations. 1978. United Nations Conference on Desertification. Round-up, Plan of Action, and Resolutions. United Nations, New York. 43 p.

2. DESERTIFICATION PROCESSES

Land degradation by man is initiated when the vegetative cover is reduced or destroyed by overgrazing, cultivation, woodcutting, vehicular traffic, improper management of irrigation water, and other activities that disturb the natural condition. If overgrazing, for example, is allowed to continue unchecked, conditions ultimately may deteriorate to the point where the land is no longer sufficiently productive to provide a livelihood for the pastoralist. This extreme degree of degradation is referred to here as very severe desertification. That condition, fortunately, occurs in only a small fraction of the world's arid regions, whereas serious but less severe desertification is widespread. Instantaneous very severe desertification is what happens when giant machines strip away the soil overlying coal beds.

Three major processes are involved in the land deterioration that we call desertification. First is the degradation of vegetative cover, second is soil erosion, and third is waterlogging and salinization. Soil compaction is a fourth process that undoubtedly is important wherever livestock and machinery pack the soil but little is known about the magnitude and areal extent of the problem. The first two processes affect the most land, by far, and are the most significant in their effort on environmental quality and on human wellbeing. Other processes include land pollution by toxic inorganic and organic chemicals, such as heavy metals and chlorinated hydrocarbons, and by industrial and municipal wastes.

The likelihood that man's land degradation activities will lead to the very severe and essentially irreversible desertification represented by mobile sand dunes and deep gullies depends, principally, upon the length of time those activities are continued (Le Houérou, 1969). A few years of abuse, followed by good management, can generally be tolerated with minimal permanent impact. There are exceptions. A single trip by a heavy truck across a sparsely vegetated area may leave tracks that persist indefinitely. And there is no way that the land disturbed by surface mining will recover rapidly if left untreated. The drier the region, the less disturbance it takes to leave a permanent scar and the more difficult it becomes to repair the damage.

PRINCIPAL PROCESSES

The four desertification processes having the most extensive impact on the biological productivity of land are degradation of the vegetative cover, soil erosion, salinization and waterlogging, and soil compaction.

Degradation of Vegetative Cover

Plant cover - on grazing and cultivated lands - serves as a direct or indirect source of human food and as protection against soil erosion. Under natural conditions, plant cover varies from sparse or non-existent in the extremely arid deserts to fairly dense in the wetter part of the semi-arid regions. The potential for damage to plants is greatest in the drier part of the arid regions, whereas the potential for recovery is greatest in the wetter areas. This means that a given percentage of destruction of vegetation in the drier areas is more difficult to reverse than the same amount of degradation in the wetter areas. In turn, the likelihood of being able to economically improve grazing lands by introducing new grazing control techniques becomes less as the climate becomes drier. The implication is that range improvement should begin on the more favorable sites, not the worst ones, if economic viability is the criterion of success. Similarly, the practice of minimum tillage to control erosion in cropland is more effective when rainfall conditions are favorable than when they are unfavorable.

Kassas (1970) described a shift of vegetational belts on the southern edge of the Sahara in the Sudan as "desert creep". The shift in vegetational belts, following overgrazing, woodcutting, erosion, and soil compaction, led to more xerophytic vegetation supplanting more mesophytic types. In the Sudan, the desert vegetation type replaces the steppe type, the steppe type replaces the savanna vegetation, and the savanna replaces the forest, each change bringing into dominance more xerophytic vegetation communities. Many of those changes have taken place within living memory. The climate has not changed but soil moisture conditions have become less favorable in the degraded areas (Le Houérou, 1976).

In southern New Mexico, York and Dick-Peddie (1969) documented a similar, presumably permanent, change in the local environment after overgrazing and erosion. They say that the increased edaphic (soil) aridity has changed the potential climax vegetation from desert grassland to desert shrub. Increased soil aridity, without any climatic change, probably is more responsible for decreasing land productivity than is generally realized. Certainly the severe water erosion that occurred centuries ago on Mediterranean slopes has altered the soil environment permanently, to all intents and purposes. The lost topsoil is gone forever, leaving a more hostile plant environment.

Range scientists evaluate the capacity of range sites to produce forage by determining range condition. The higher the range condition rating, the closer the range approaches its potential for forage production. Range trend evaluations are made at two or more time periods to estimate whether the range condition is getting better or worse.

Water and Wind Erosion

Water erosion is the dominant type of erosion in the arid regions, in the extremely arid deserts as in the semi-arid grasslands. Wind erosion carves strange rock formations that remind one of ships or birds or sphinxes, and piles sand in beautiful geometric shapes, but water erosion does most of the work of changing the land surface.

Erosion may be geologic (natural) or accelerated (due to man's activities). In desertification, the concern is with accelerated erosion, on the premise that geologic erosion is very difficult if not impossible to control for any extended period unless truly heroic measures are employed. Accelerated erosion, since it is due to man's abuse of the land, can be controlled by altering man's activities. That does not imply that the control can be done economically or that the original plant and soil conditions can be restored. Control is cheapest and most effective in restoring maximum soil productivity if it is initiated before significant degradation has occurred.

Differentiating between geologic and accelerated erosion, in order to determine the degree to which erosion can be reduced, can be very difficult. A good example is the very shallow soil found on shale deposits in the central Rocky Mountain states of the United States. Shale is, under the best of circumstances, highly susceptible to water erosion. Disturbing the sparse vegetative cover found on shale soils in the arid regions undoubtedly increases runoff and erosion, but not necessarily by very much. The muddy color of the intermittent and permanent streams may be largely due to natural conditions rather than to the effects of man. Difficult though the distinction may be to make, it must be attempted if desertification effects are to be analyzed properly.

A similar problem of separating geologic from accelerated erosion can be seen in studying mobile sand dunes. Stabilization of moving sand dunes can best be accomplished by revegetating them. If the climate is such that re-vegetation is successful without permanent irrigation, the implication is that the mobile dunes are the result of man's intervention. If on the other hand, re-vegetating is not a viable control measure, the wind erosion is presumed to be geologic.

Dune encroachment on oases in extremely arid climates

can be construed as a special case of desertification.
In this instance, where dune movement is a natural
phenomenon unaffected by man's activities, the mistake
made by man is to put his dwellings and fields directly
in the path of the invading dune. Under that circumstance,
the battle against dune encroachment will be never-ending,
with the dune system likely to be the ultimate winner.
Numerous oases in the Sahara are threatened in this manner
or have already succumbed to the forces of nature.

Severity of water and wind erosion is determined by
estimating or measuring the loss of soil and, in the case
of wind erosion, the amount of deposition that has taken
place. Actual measurements of soil loss are rarely made
because the task is so difficult under field conditions.
Instead, estimates are made from calculations such as the
Universal Soil Loss Equation for water erosion (Wischmeier
and Smith, 1978) and the Wind Erosion Equation for wind
erosion (Woodruff and Siddoway, 1965) or by visual or
photographic inspection of the land area.

Salinization and Waterlogging

Most commonly, salinization and waterlogging of
irrigated land go together, with waterlogging leading to
salinization. However, salinization can also be the
result of applying saline irrigation water to slowly
permeable soils or applying too little water to hold soil
salinity to a minimum. In both irrigated and dryland
agriculture, salts can also accumulate on the soil surface
as the result of lateral seepage of water and dissolved
salts.

Salinization is a greater threat to crop production than is waterlogging. In the absence of salt effects, a water table probably must be within 50 cm of the soil surface before crop yields will be materially affected. In the presence of toxic salts, a water table within two meters of the surface usually poses a problem. Worldwide, the damage done by high water tables, alone, is relatively small in irrigated areas.

Guidelines for predicting the effect of different levels of soil salinity on crop yields have been established (Salinity Laboratory Staff, 1954). The soil tests utilize the electrical conductivity of a saturation extract to measure salinity. If a sodic problem is present, its severity is assessed by determining exchangeable sodium percentage. Irrigation water quality can be evaluated by using FAO guidelines (Ayers and Westcot, 1976).

Soil Compaction

Arid region soils are especially prone to soil compaction because they usually are low in organic matter, are not generally subject to severe freezing and thawing, and are frequently dry to a considerable depth. Low levels of organic matter facilitate compaction; freezing and thawing reduce compaction. Dehydration promotes compaction, although the most common causes of significant compaction are the pressure of tillage machinery, the pounding of livestock hoofs, and the dispersion and packing effect of raindrops striking the soil.

Surface soil compaction is called soil crusting, which is usually caused by raindrops, animal hoofs, dehydration of the soil, and tractors. The pressure exerted by tractors may pack the soil to a depth of 30 or more centimeters, whereas tillage implements such as plows compact a subsoil layer at the point where the implement presses on the soil. Those compact subsoil layers are called plow pans or plow soles.

Soil crusting is widespread on grazing land and on cultivated land. Deeper soil compaction is found on cultivated land, both dryland and irrigated.

The magnitude of the soil compaction problem is unknown but there is general agreement that it is common on cultivated land where mechanization is practiced. Surface soil crusting must also be common on grazing land, judging by the numerous references to its occurrence. One author has estimated that soil compaction in row crops in the United States reduces yields about 10 percent in the southern states where soils seldom freeze below 25 cm (Gill, 1971).

Soil compaction can be measured by determining modulus of rupture, measuring soil density, or pushing a penetrometer into the soil, or using indirect tests. Interpreting soil density or strength data is difficult because the tests are subject to numerous variables. It usually is necessary to compare compaction data with those taken at other times or in other layers above and below the layer in question. As a consequence, few tests are made except at research sites.

DESERTIFICATION AND LAND USE

Desertification processes vary with land use although they can all be grouped under the two categories of vegetation and soil degradation. Three major types of land use are practiced in the arid regions: grazing, dryland farming, and irrigated agriculture. Mining, tourism, recreation, and urban development have a less extensive - but important - impact.

GRAZING LAND

The desertification process on grazing land has three phases: 1) perennial grasses are replaced by annual grasses and forbs, usually followed by an increase in the density of shrub and tree types, 2) an absolute decrease in the amount of plant cover of all types occurs, leaving soil less protected from the impact of livestock hooves and the erosive force of wind and water, and 3) soil crusting caused by animals trampling the bare ground leads to increased runoff and greater wind and water erosion.

Overgrazing

The degradation process that is the most common in rangelands begins with livestock overgrazing the desirable perennials and annuals, leaving the way open for an invasion or increase of less palatable grasses, forbs, and shrubs and of poisonous plants, as competition by perennial grasses is reduced (U.S.S.R., 1977; Department of the Environment, 1977; Bernus, 1977; Heady and Bartolome, 1977). Total biomass may, in the early stages of desertification, be increased, but with a less desirable mixture of species. Ultimately, after continued overgrazing and trampling have been followed by increased runoff and accelerated erosion, total plant production decreases. In much of the developing world, woodcutting for fuel and other purposes contributes significantly to further deterioration of environmental conditions by removing shrubs and trees that provide protection to the soil (Eckholm, 1975).

Desertification of grazing lands becomes extreme when destruction of vegetation leads to the formation of massive gullies and sand dunes. That kind of land degradation constitutes irreversible - from the economic point of view - desertification. Restoration of such land to even a moderately productive state becomes prohibitively expensive for pastoralists, and natural recovery will take decades or centuries to occur.

Use of the term "desert encroachment" or "the advancing desert" to label desertification of grazing lands can give the impression that land deterioration spreads like a wave from a desert center. Such is not the case, usually. Rather, particular range areas are abused in a spotty pattern, with good rangeland interspersed among degraded areas. The spots may coalesce, in time, if land abuse spreads, but the starting point can be anywhere. Watering points commonly are the first places around which land degradation occurs. It is there that livestock concentrate, putting heavy grazing pressure on the nearby range. Figures on the average number of livestock grazing a large area are misleading with respect to grazing pressure. Animals rarely are scattered evenly over grazing lands. Rather, they are concentrated in herds which may overgraze an area near a pond while undergrazing more remote areas.

There are few places in the world where grazing arid rangelands by domestic livestock has been anything but destructive. Nomads probably came the closest to maintaining an ecological balance before the availability of veterinary services for their animals and the drilling of deep wells allowed them to increase livestock numbers and to intensify grazing pressures. In modern times, practicing good range management has, paradoxically become more difficult even as research has shown how it can be done better. As human and livestock populations increase, the ability and willingness of pastoralists to reduce livestock numbers - usually a first step in controlling overgrazing and in restoring land productivity - is lessened because of survival requirements. Even when pastoralists agree with the need to manage their ranges better and accept the practices that are recommended, they may be unable to adopt them if they must rely solely upon their own resources for survival during the period of change.

Lack of knowledge about rotational grazing, deferred grazing, plant response to light and heavy grazing, and other management factors was responsible for at least part of the widespread overgrazing of United States rangelands that occurred during the 19th and early 20th centuries. It was not until well into the 20th century that range scientists were able to recommend management practices that would enable ranchers to obtain high meat production while maintaining the natural resource base. Only in recent years has range management become a major concern in many countries where range livestock production is a basic element of the society. And in some countries it still receives little attention. One reason is that pastoralism in the arid regions is an extensive rather than intensive type of land use. It can never compete with irrigated land in economic returns and, thus, is less attractive to planners and developers. Furthermore, there are no "quick technological fixes" to solve the problems of overgrazing; vegetation responses to management are usually slow and dependent upon the vagaries of weather. Because of the dependence on rainfall, manipulation of range vegetation is difficult in the drier regions and easier in the wetter, and management options become more restricted as rainfall decreases.

Woodcutting

Woody plant species on grazing lands are cut for fuelwood, buildings, implements, "bush fencing" to protect livestock at night, and for other purposes. Their removal exposes the soil to accelerated erosion and increased runoff. Wood is the sole or major source of fuel for cooking and heating in the poor less developed countries. Kerosene looked to be a promising alternative fuel until oil prices increased drastically in the mid-1970's. Solar cookers and heaters and biogas generators offer promise, under some circumstances, but their cost is too high for poor people to afford. Improved wood stoves and charcoal ovens allow wood to be burned more efficiently but, as yet, their use is limited.

Consequently, woodcutting continues to contribute to environmental deterioration, and will for the foreseeable future.

Reforestation or afforestation is the key, at present, to alleviating the destructive effects of woodcutting. In all but a very few instances, there is no reason why shrubs and trees cannot be established again where they grew previously. Fast-growing and drought tolerant trees such as eucalyptus, casuarina, and mesquite (Prosopis spp.) are available for use in semi-arid regions. Techniques for seedling establishment have been developed that are usually successful (Weber, 1977). The reforestation problem is a social one, not a technical one. Given a supply of adapted species, there must be adequate incentive for the target population to plant the seedlings and to protect them while they are young. Usually, that incentive is missing.

Fire

Fire can be used to improve or to degrade rangelands. The difference is in when the burning is done and how the land is managed subsequently. In the semi-arid and sub-humid regions, fire can control the spread of undesirable shrubs and trees and encourage the growth of palatable grasses. If done at the wrong time, it can also increase soil erosion by denuding the land just when protection is needed from pounding raindrops. How firing is done determines whether its effects are good or bad (Wright and Britton, 1976).

DESERTIFICATION PROCESSES

Soil Compaction

The kind of soil compaction occurring in rangelands is surface soil crusting. Crusting is the result of packing surface soil particles together. It is a natural phenomenon in the extremely arid and arid regions where low rainfall makes the vegetative cover sparse. In the better rangelands, crusting is not an important problem until overgrazing or woodcutting or fire reduce surface litter and expose the soil. The beating action of raindrops on the bare soil and the pressure exerted by the hoofs of livestock are the major causes of surface crusting. The effect of crusts is to reduce infiltration rates and increase runoff, with the dual result of less moisture entering the soil for plant use and more water erosion occurring.

DRYLAND CROPPING

The desertification process that goes on when dryland cropping begins is straightforward: the native vegetative cover is destroyed by cultivation or burning, exposing the soil to accelerated wind and water erosion. Some degree of protection - usually small - is afforded the soil by the growing crop once it has been established. That protection is lost as soon as or shortly after the crop is harvested unless the plant stalks are left intact to reduce the impact of wind and water. Even when the stalks are left standing, the common practice of grazing stubble soon reduces or eliminates the remaining plant cover. As a consequence, the soil is unprotected for periods of several months each year under annual cropping or for longer periods if clean fallowing is practiced as a means to conserve moisture or to mineralize organic nitrogen in the soil. Tractors and heavy tillage implements increase subsoil compaction, and exposing the unprotected soil to raindrop impact increases surface crust formation. Compaction and crusting reduce soil permeability and increase runoff and erosion.

Erosion by either or both wind and water, in the early stages of cultivation, is generally unnoticed or is considered to be unimportant. There are exceptions, of course, and they commonly are associated with unusually hard rains that may cause gullies to form almost overnight or with severe droughts and strong winds that strip away the exposed soil and pile it up in hummocks or dunes. Those exceptions illustrate dramatically the destructive power of wind and water erosion, but gullies and dunes are, in most cases, only the end products of long-continued neglect of the land (Instituto de Investigaciones Agropecuarias, 1977; Stallings, 1957; Central Arid Zone Research Institute, 1977).

Effects of Erosion

Wind and water erosion are the greatest threats to continued dryland cropping and are the principal processes by which desertification occurs on such lands. Wind and water erosion have much in common in the way they affect the land. Sheet erosion is the first step in the desertification process. At this stage, the finer soil particles of silt and clay, along with organic matter attached to them, are washed or blown away. For a few years, there may be no observable effect on the soil or on the crops of that removal.

In time, with continued sheet erosion, the removal of the more fertile topsoil and the exposure of less productive subsoil leads to declining yields unless nutrient supplies are replenished. The decline in nutrient availability - principally nitrogen - may not be noticed in relatively dry years when crop demands for nutrients are low. In years when moisture conditions are favorable, however, nutrient supplies become inadequate to meet crop demands, and yields suffer. Over a long period of sheet erosion there is a gradual reduction in yields if cropping practices remain the same. At some point, little or no further reduction occurs and yields tend to remain at a low level.

The second - and more severe - stage of desertification occurs when gullies and dunes appear. Gullies represent true land destruction: they are completely useless for any kind of conventional agriculture. Small gullies may be filled in or smoothed out at considerable expense. Large gullies scar the land forever, for all practical purposes. Sand hummocks and small dunes can sometimes either be cultivated as they are or leveled to restore a reasonably smooth surface. Large barren dunes, however, are useless for cropping purposes. Gullies begin on the lower slopes and move upward as erosion continues; sand dunes form downwind from the point of soil removal and move in the direction the wind blows.

Wind and water erosion are similar, in many respects, in their effect on the land. The physical laws governing soil movement by wind or water are the same since both of the transporting media behave as fluids. Sheet removal of soil particles is initiated at the beginning of both wind and water erosion. In the case of water erosion, hilltops are affected most by soil loss, side slopes are less affected, and deposition of coarse textured material occurs on the lower slopes where water velocity is reduced. For wind erosion, the greatest soil removal occurs near the point where erosion begins; deposition occurs downwind wherever an obstacle of some kind reduces wind velocity. In between, removal and deposition of coarse particles are intermixed; the fine suspended particles are carried far away, just as they are in water erosion. Dust from Texas has been collected thousands of kilometers away over the Atlantic Ocean, as has dust from Sahara windstorms, and such events are not unusual.

Wind erosion is different from water erosion in one important effect it has on plants: sand blasting. Moving sand can be highly destructive to plants by shredding leaves through abrasion and, in extreme cases, by killing entire plants. Sand blasting injury to seedlings can make it necessary to replant fields several times in order to obtain a satisfactory stand.

Saline Seeps

Development of saline seeps has become recognized as a significant source of soil degradation in dryland farming areas of certain parts of North America and Australia (Vander Pluym, 1978). The saline seep problem is a serious one in the wheat growing regions of the northern Great Plains of Canada and the United States and in the Australian states of New South Wales, Victoria, and Western Australia, and it continues to get worse. Saline seeps would be expected to occur wherever dryland small grain crops are produced on soils underlain by slightly to strongly saline parent material and a slowly permeable soil layer and occurring on slopes. The man-made problem is the result of a reduction in evapotranspiration when deep-rooted native vegetation is replaced by shallow-rooted crops that transpire less water. Replacement of trees by wheat is an example. Fallowing exacerbates the problem. Increased percolation of rainfall through the soil carries soil salts - if present - to the surface in downslope areas, causing saline seepage spots to appear.

Soil Compaction

Surface soil crusting and subsurface compaction are both present in dryland farming fields. Surface crusting is the resultant of a lowered soil organic matter content in cultivated soils, the impact of raindrops, and the pressure of machinery. Subsoil compaction is caused by heavy machinery and by repeated tillage at about the same depth.

Surface crusting impedes seedling emergence, increases runoff, and contributes to increased wind erosion by smoothing the soil surface. Subsurface compaction restricts water and root penetration and increases runoff.

Formation of dense crusts on the surface of agricultural soils after a rain or irrigation is common. The problem is a critical one in vegetable production, where even a fairly thin crust can prevent emergence of tiny seedlings (Hoyle, 1983). Crusting occurs sporadically, depending upon a number of factors such as intensity and duration of rain. No practical method of preventing crusting in cultivated soils has yet been devised. Various synthetic organic materials, of which Krilium was the most widely publicized, have been tested. Many of the polymers have been effective; the limiting factor is cost.

IRRIGATED AGRICULTURE

Salinization and waterlogging have been the curse of irrigation agriculture since water was first diverted from rivers crossing the arid regions. Wherever irrigation is practiced in the dry regions, the threat or the reality of salinization is always present. While waterlogging, alone, can reduce crop yields severely, the salinization that usually accompanies waterlogging is the greater threat. Owing to high surface evaporation rates and the upward movement of water and dissolved salts in a waterlogged soil, salts may accumulate on the soil surface in toxic amounts even though the salt content of the irrigation water is low. The classic example of that progression of salt accumulation is found in the Mesopotamian Plain of Iraq but similar events have occurred in the Indus Valley of Pakistan, along the Colorado River in Mexico and the United States, in the Nile Delta of Egypt, and elsewhere. The Iraq and Pakistan situations have been described in detail in documents prepared for the United Nations Conference on Desertification (Dougrameji and Clor, 1977; Irrigation, Drainage, and Flood Control Research Council, 1977).

In relatively few - but locally important - instances, salt accumulations are due to the application of high-salt irrigation water to slowly permeable soils. That type of problem is especially serious in much of Tunisia and in the oases of Algeria and the Middle East. Under conditions of restricted soil permeability and little or no leaching, salts accumulate when the applied water is evaporated or taken up by plants. The process is essentially the same as that responsible for the formation of natural salt lakes, salinas, sebkhas, and salares in non-irrigated areas.

Waterlogging

Waterlogging adversely affects crop growth by reducing soil aeration. Rice is the notable example of the exceptional crop that can grow in soils saturated with water. For most crops, the depth at which free water occurs in the soil is the limit for root growth. If the water table is 40 centimeters below the soil surface, the effective crop root zone is only the 40 centimeters of unsaturated soil between the surface and the water table. Roots are unable to survive in unaerated soil without special structural adaptations. Waterlogging and associated salinity problems are minimal when the water table is more than two meters below the surface.

The high water tables that cause trouble in irrigated areas are almost always the result of irrigation water percolating through the soil and accumulating above a slowly permeable layer in the subsoil. In some cases, the lateral movement of groundwater from higher lying land is responsible for a rising water table in the lower land. The solution to the problem, in both cases, is to reduce the amount of irrigation water applied or to provide artificial drainage through open or closed drainage systems.

Salinization

Desertification through salinization is due to the accumulation of soluble salts in the soil to the point where plant growth is impaired. Salts usually are found in largest amounts on or near the soil surface. In extreme cases, the ground may be covered with a white salt crust. More commonly, the salts are distributed throughout the soil profile, with higher concentrations at the top or bottom, depending upon the amount of leaching that occurs and the time of sampling for salinity measurements. Samples taken immediately before an irrigation are likely to show the greatest salinity near the surface whereas samples taken shortly after an irrigation tend to show low salinity values for the surface soil. Leaching moves the salts downward, evaporation moves them upward.

Since crop plants differ greatly in salt tolerance, the effect of a given level of salt may range from no effect to a very severe effect. Salt tolerant barley and sugar beets may not be affected at all by salinity levels that would prevent growth of salt sensitive carrots and beans. In addition to salinity response differences due to plant tolerance, there may also be a differential response due to irrigation practices, planting techniques, and weather conditions. As soil salinity levels increase, the effects change from a slight decrease in plant growth (with no visual indication of trouble) to a greater growth reduction (with leaf tip scorching) and finally to the worst case where the land is barren because no plants can become established. The generally spotty distribution of salt affected soils means that one part of a field may show signs of salt toxicity while another part is unaffected.

Salinization sometimes leads to another kind of problem, that of accumulation of adsorbed sodium on clay particles. When the level of adsorbed sodium is high and the salinity is low, soil structure becomes unstable and dispersion of clay particles occurs. The result is the formation of a sodic or black alkali soil in which permeability is reduced due to plugging of soil pores by the dispersed clay. In addition to the harmful effects on crop plants of the reduced permeability and the resulting poor aeration, sodic soils upset the nutritional balance of plants by virtue of their excess adsorbed sodium and high alkalinity.

Soil Compaction and Crusting

The pressure exerted by tractors and tillage implements tends to form a compact layer in the soil at the bottom of the tilled layer (Doneen and Henderson, 1953). Root and water penetration into these plow pans or tillage pans is restricted. Man-induced compacted layers are common in fine textured irrigated soils, especially. They also are widespread in dryland soils where mechanization is practiced. Soil compaction leads to increased surface water runoff, reduced plant growth, and greater susceptibility of crops to drought. Crusting of surface soils occurs when raindrops beat against bare soil or when animal hoofs and machinery break down soil aggregates. Once the soil particles have become dispersed, they flow together during rains and form a crust that increases runoff and erosion. If the crusts are hard enough, seedling emergence is impaired, sometimes to the point where fields must be planted again in order to achieve a satisfactory stand.

MINING

Surface strip mining brings about immediate and extreme desertification in a spectacular fashion. Suddenly there is a large and barren hole in the ground that may ultimately be five to ten meters deep or - as in the case of open pit copper mines - a hundred or more meters deep. The overburden of soil that was removed to get access to the subsurface minerals may be lined up in parallel rows along one side of the excavation - as in coal mining - or spread out nearby. Somewhere there will be a processing plant where waste material is separated and discarded into a slag dump. Usually the slag dump and the overburden are sterile and completely devoid of vegetation, due to the strong acidity, lack of nutrient elements, and presence of toxic substances in the discarded rock and debris. Slag dumps are notorious in the arid regions as sources of wind-blown heavy metals that have adverse effects on downwind vegetation. Their dry and barren surfaces provide ideal conditions for wind erosion.

Underground mining has much less detrimental impact on the environment. Vegetation around the mine or well entrance is destroyed, commonly, but the slag piles do the more extensive damage. Oil and gas wells and underground mines usually are readily discernible from the air because of the very severe desertification in the immediate vicinity of the well or mine. Mining also has an indirect effect on land degradation in that roads constructed to lead to the mines generally are unimproved. They encourage wind and water erosion along their length and can serve as the starting point for accelerated erosion on adjoining land.

The total land area desertified by mining is miniscule in comparison to that affected by grazing and cultivation, but the visual impact is greater. It is hard to ignore an immense and empty hole in the ground and piles of sterile earth covering dozens of hectares of land.

TOURISM AND RECREATION

Arid regions have been promoted as ideal places for tourists to go to enjoy the sun and scenery. Oases such as Las Vegas in the United States, Gafsa in Tunisia, Arica in Chile, and Tamanrasset in Algeria have special attraction for people from the cold climates. With tourism comes the threat of land destruction when too many people congregate in a small area. By and large, tourism is not destructive of the natural environment until people get into off-road vehicles that give them access to places where there are no roads. Then the potential for instant desertification is greatly increased.

Off-road vehicle use for recreation has caused much destruction in the California desert of the United States

(U.S. Department of the Interior, 1968) and elsewhere. Four-wheel drive vehicles driven carelessly can - and do - knock down shrubs and leave ruts in the soil that increases susceptibility to wind and water erosion. Parking and camping areas resemble livestock watering points in the destruction of vegetation that occurs there. Cross country motorcycle races are hazardous to vegetation as well as to the life and limb of the racers.

WOODCUTTING

Cutting of wood for direct burning as firewood or for the manufacture of charcoal provides fuel for as much as ninety percent of the people in poor countries (Eckholm, 1975). Wood cutting has increased in recent years as the population and the cost of kerosene increased. The magnitude of the effect woodcutting has had on vegetative cover around cities is indicated by the estimate (Delwaulle and Roederer, 1973) that deforestation to a radius of 40 to 50 kilometers around Niamey, Niger was nearly total because of the need for firewood.

In the absence of other fuel sources, woodcutting can be expected to continue for the foreseeable future, briging with it widespread desertification as the shrub and tree cover is eliminated. Attempts to re-establish trees to reduce soil erosion have often been thwarted by the almost immediate removal of the trees by local people seeking firewood. Given the shortage of fuel, the costliness of kerosene, and the convenience of using wood and charcoal - when available - for cooking and heating, it will be difficult indeed to stop indiscriminate cutting of shrubs and trees.

Overexploitation of forest resources to make charcoal for iron smelting has been linked to the development of a sophisticated and highly efficient smelting process in Tanzania as much as 1500 years ago (Schmidt and Avery, 1978).

Similar heavy demands on wood for smelting in the arid Great Basin of the United States in the 1800's was responsible for denuding land of trees for dozens of kilometers around the numerous small smelters that sprang up there. Thus, desertification due to woodcutting is not new even in places where the population density was relatively low.

URBAN DEVELOPMENT

Small towns in pastoral and mining areas frequently are unattractive because land in and around the town is badly degraded. Houses tend to be widely spaced, the land between the houses in untended, water is in short supply, and soil surfaces are bare and dusty. Mining towns are notorious for their blighted look, with environmental degradation seemingly everywhere. Large numbers of people converging on a small community over a short period of time, as frequently happens around newly opened mines, has a devastating effect. Once the vegetation and soil have been severely disturbed, restoration becomes difficult, at best. Coupled with the fact that the population usually is more or less transient, mining communities in arid regions have special problems in controlling desertification.

Land surrounding the towns also suffers degradation from grazing pressure, land excavation, waste disposal, recreation, woodcutting, and incidental activities. How far the adverse impact is felt depends upon the size of the community and its economic reliance on local resources.

CONCLUSIONS

Desertification on the 224,000,000 hectares of rainfed cropland and the 3,750,000,000 hectares of rangeland in the arid regions is primarily a problem of loss of vegetative cover. Without a good plant cover, the productivity of the land for animal products is lessened and the susceptibility to wind and water erosion, soil compaction, and soil crusting is increased. No economically better substitute has yet been found for living plants or crop stubble as the means of stopping accelerated erosion. Asphalt, plastic, gravel, soil conditioners, and concrete will do the job, but at a high price.

The 126,000,000 hectares of arid irrigated land are continually threatened with waterlogging and salinization. Prophets of doom contend that those twin evils assure that irrigation agriculture cannot survive indefinitely. It has lasted several thousand years in the Nile Valley because periodic flooding occurred. With the present complete control of the river in Egypt, it is only a matter of time before the agricultural economy collapses - or so it is said. Optimists believe that man has it within his power to minimize salinity and high water table problems, if he will use that power wisely. Experience in olden Iraq demonstrates that controlled irrigation can be successful indefinitely if constant vigilance is used in the management of the system.

REFERENCES

Ayers, R. S., and D. W. Westcot. 1976. Water Quality for Agriculture. Food and Agriculture Organization, Irrigation and Drainage Paper 29, Rome, Italy. 97 p.

Bernus, Edmond. 1977. Case study on desertification. The Eghazer and Azawak Region. Niger. United Nations Conference on Desertification A/CONF. 74/14. 111 p.

Central Arid Zone Research Institute. 1977. Case study on desertification. Luni Development Block. India. United Nations Conference on Desertification. A/CONF. 74/11. 66 p.

Delwaulle, J. C., and Y. Roederer. 1973. Le bois de feu á Niamey. Bois et Forêts des Tropiques, No. 152, pp. 55-60.

Department of the Environment. 1977. Case study on desertification. Iran: Turan. Government of Iran, Tehran. May 1977. 97 p.

Doneen, L. D., and D. W. Henderson. 1953. Compaction of irrigated soils by tractors. Agricultural Engineering 34: 94-95.

Dougrameji, J. S., and M. A. Clor. 1977. Case study on desertification. Greater Mussayeb Project. Iraq. United Nations Conference on Desertification. A/CONF. 74/10. 101 p.

Eckholm, Erik P. 1975. The Other Energy Crisis: Firewood. Worldwatch Paper 1. Worldwatch Institute, Washington, D.C. 22 p.

Gill, William R. 1971. Economic assessment of soil compaction. In Barnes, K. K., W. M. Carleton, H. M. Taylor, R. I. Throckmorton, and G. E. Vander Berg (editors), Compaction of Agricultural Soils. American Society of Agricultural Engineers, St. Joseph, Michigan. pp. 431-458.

Heady, Harold F., and James Bartolome. 1977. The Vale Rangeland Rehabilitation Program: The Desert Repaired in Southeastern Oregon. USDA Forest Service Resource Bulletin PNW-70. Pacific Northwest Forest and Range Experiment Station, Portland, Oregon. 139 p.

Hoyle, Burton J. 1983. Crust control aid seedling emergence. California Agriculture Volume 37, No. 1 and 2, p. 25-26.

Instituto de Investigaciones Agropecuarias. 1977. Case study of desertification. Region of Combarbala. Chile. United Nations Conference on Desertification. A/CONF. 74/9. 162 p.

Irrigation, Drainage, and Flood Control Research Council. 1977. Case study on desertification. Mona Reclamation Experimental Project. Pakistan. United Nations Conference on Desertification. A/CONF. 120 p.

Kassas, M. 1970. Desertification versus potential for recovery in circum-Saharan territories. In H. E. Dregne (editor), Arid Lands in Transition. American Association for the Advancement of Science, Publication No. 90, Washington, D.C. pp. 123-142.

Le Houérou, Henri-Noel. 1969. Principes, Methodes et Techniques d'Amelioration Pastorale et Fourragere en Tunisie. Food and Agriculture Organization of the United Nations, Etude No. 2, Paturages et Cultures Fourrageres. Rome, Italy. 291 p.

Le Houérou, H. N. 1976. Can desertification be halted? In Conservation in Arid and Semi-Arid Zones. FAO Conservation Guide 3. Rome, Italy. pp. 1-15.

Salinity Laboratory Staff. 1954. Diagnosis and Improvement of Saline and Alkali Soils. U.S. Department of Agriculture, Agriculture Handbook No. 60, Washington, D.C. 160 p.

Schmidt, Peter, and Donald H. Avery. 1978. Complex iron smelting and prehistoric culture in Tanzania. Science 201: 1085-1089.

Stallings, J. H. 1957. Soil Conservation. Prentice-Hall, Inc., Englewood Cliffs, New Jersey. 575 p.

U.S. Department of the Interior. 1968. The California Desert. Bureau of Land Management, U.S. Department of the Interior. Sacramento, California. 387 p.

U.S.S.R. 1977. Golodnaya Steppe. An Associated Case Study. United Nations Conference on Desertification. A/CONF. 74/23. 76 p.

Vander Pluym, H. S. A. 1978. Dryland-Saline-Seep control. Alberta Agriculture, Agriculture Center, Lethbridge, Alberta, Canada T1J 4C7. 315 p.

Weber, Fred R. 1977. Reforestation in Arid Lands. Volunteers in Technical Assistance, Washington, D.C. 248 p.

Wischmeier, W. H., and D. D. Smith. 1978. Predicting Rainfall Erosion Losses: A Guide to Conservation Planning. U.S. Department of Agriculture Agriculture Handbook No. 537, Washington, D.C. 58 p.

Woodruff, N. P., and F. H. Siddoway. 1965. A wind erosion equation. Soil Science Society of America Proceedings 29: 602-608.

Wright, H. A., and C. M. Britton. 1976. Effects of fire on vegetation in western rangeland communities. In The Role of Prescribed Burning in Western Range and Woodland Ecosystems. Utah State University, Logan, Utah. pp. 49-53.

York, John C., and William A. Dick-Peddie. 1969. Vegetation changes in southern New Mexico during the past hundred years. In William G. McGinnies and Bram J. Goldman (editors), Arid Lands in Perspective, University of Arizona Press, Tucson, pp. 156-166.

3. DESERTIFICATION INDICATORS

Desertification begins in a manner that seldom arouses concern in its early stages. A few bare spots in a grassland, a little more sediment runoff than in previous years, a few small hummocks of sand appearing around some shrubs, or a white crust on part of one row in an irrigated field does not appear serious, and sometimes it isn't. But all too frequently those first indications of land degradation are followed by worsening conditions that soon assume significant proportions.

Studies on indicators of desertification have been confined almost exclusively to rangelands. Those studies were not directed toward finding indicators of incipient or extreme desertification (desertification was not a widely used term until recently). Instead, they were intended to provide data on range vegetation conditions that would permit decisions to be made about the need for changes in management practices. Knowing the potential productivity, the present condition, and the condition trend (whether it is getting better or worse or staying about the same) of range sites over time means that the degree and rate of desertification can be determined since range deterioration is one form of desertification. Similar determinations can be made for other land uses, but that has not been done to anywhere near the degree that it has for rangelands.

Indicators of desertification must be indicators of changes that have occurred. Indicators should, desirably,

DESERTIFICATION INDICATORS

be quantitative, sensitive to small changes, easy to measure and small in number. Direct or indirect measurements may be used. Those physical, biological, or sociological factors that integrate a number of other factors are especially useful. An example is vegetation degradation in a grazing area, as measured by meat production. Changes in meat production can be a useful indicator of changes in the quantity and quality of plant biomass due to overgrazing, soil erosion, salinization, etc. if factors such as disease and drought do not complicate the interpretation. While the measurement of desertification indicators should be simple enough for a non-specialist to carry out, the interpretation of the significance of the condition measured must be done by an experienced individual.

Desertification indicators can serve two purposes: they can be used to estimate how much land degradation has occurred in the past and what effect current management practices are having on the ecosystem.

GRAZING LANDS

Range scientists in the United States recognized early that recommendations for appropriate range management practices depended upon knowing what the current condition was relative to what it could be with improved management. That recognition led to the search for indicators of range conditions and trends. The closer a range site is to its maximum potential for forage production, the better is the range condition. The maximum potential depends upon the soil and climate. Range condition is usually described as excellent, good, fair, or poor, relative to the potential.

The three federal agencies most heavily involved in range management on federal, state, and private lands in the United States are the Soil Conservation Service, the Forest Service, and the Bureau of Land Management. Together, they are the major source of management advice on the approximately 320 million hectares of rangelands in the 17 western states. Each of the three evaluates range condition differently. The Soil Conservation Service bases its evaluation on plant groups; the other two agencies consider both plant and soil factors (Gay, 1965).

Range Condition

In the Soil Conservation Service range condition classification system, plants are placed in three groups according to their response to grazing. Decreaser plants are those desirable forage plants in the native (climax) vegetation that gradually decrease in amount as grazing pressure increases. Increaser plants are the second-choice plants in the climax vegetation. They are less heavily grazed, at first, than the more palatable and more accessible decreasers and, therefore, increase in proportion to the decreasers. As heavy grazing continues, the increasers decline in productivity. Invader plants are undesirable shrubs, grasses, and forbs that invade the rangeland as the decreaser and increaser plants die. The range condition rating for a particular site is a function of the relative amount of decreasers, increasers, and invaders present on a volume or weight basis (Table 3.1). A knowledge of the climax vegetation at a site is necessary in order to estimate how much change has been brought about by grazing. Susceptibility to soil erosion is greatest when invader plants dominate the community.

TABLE 3.1. Range condition classes, U.S. Soil Conservation Service

Condition	Percent of Maximum	Vegetation groups
Excellent	75-100	Mostly decreasers
Good	50-75	Mostly increasers, some decreasers
Fair	25-50	Mostly increasers, few decreasers, some invaders
Poor	0-25	Mostly invaders

The Forest Service range condition classification evaluates species composition - as does the Soil Conservation Service system - but also includes an evaluation of soil condition, amount of vegetative cover, and plant vigor. The Bureau of Land Management classification rates forage species composition, density, vigor, and reproduction as well as a few soil factors related to erodibility.

Plant ecologists have found the concept of climax vegetation useful in analyzing plant succession. Yet, in regions where livestock grazing has been carried on for hundreds or thousands of years, the climax concept may have little significance. Continued grazing may have altered the ecosystem to the place where identification of the original (before domestic livestock were introduced) vegetation is difficult or impossible and may even be meaningless as a guide to present grazing potential (Martin, 1975). Introduction of improved or better adapted exotic species can raise the production ceiling above what the climax vegetation is capable of. This is especially true when fertilizer application brings in a new variable, that of the ability of a plant to respond to fertilizers. Native grasses may be less able - by reason of their genetic constitution - to respond to added plant nutrients than their introduced relatives.

Range condition evaluations should give weight to soil conditions as well as plant conditions. There are at least two good reasons for doing so. One is to assist range managers in determining the ability of a range to recover rapidly from poor management. A second is to enable the manager to know where he must be particularly careful to avoid exposing the soil to erosive forces. Once the soil has been eroded down to the subsoil or is gullied, the soil environment is changed completely, for the worse. This means that combatting desertification is much more difficult on a severely eroded range in poor vegetative condition than on a slightly eroded range that also is in poor vegetative condition.

Range condition class shows how much - if any - desertification has occurred in the past. It does not tell whether conditions are changing, although a comparison of two range condition evaluations over a period of several years would show what changes had occurred.

DESERTIFICATION INDICATORS

Range Condition Trends

Range condition changes are likely to occur - if they do - over a period of years. Periodic estimates of trends in range condition can serve as an early warning system and be of immediate use to range managers. Trends are indicated by such things as changes in plant vigor, rate of plant reproduction, survival of seedlings, and the amount of plant residue (litter) retained on the soil surface. Trend evaluation is useful for determining appropriate adjustments in stocking rate when the range is underutilized or before any detrimental changes become serious.

Range condition trend indicators are also current desertification rate indicators. In combatting desertification, knowing whether current management practices are worsening or bettering the situation is sometimes more important than knowing how much desertification has occurred in the past.

RAINFED CROPLANDS

Dryland farming is, by its very nature, a type of land use that poses serious problems for land conservation. Plant growth is less than in the humid regions and frequently varies greatly from year to year in response to rainfall fluctuations, the soil usually lies bare and subject to wind and water erosion for several months each year, crop selection is restricted to those crops having a degree of drought tolerance, rotation of grain crops with soil-building grasses and legumes generally is uneconomic, and fertilizer applications seldom are profitable except in the wetter areas and in climatic regions where wheat is produced during a cool wet season. All of these factors operate to make dryland farming a risky enterprise. The risk is greater in the summer rainfall areas and lesser in the winter rainfall zones because annual ranges in precipitation and storm intensity tend to be less in the latter regions.

Soil erosion is the number one desertification problem in the rainfed cultivated lands. It follows, then, that the most important direct indicators of desertification are changes in soil characteristics. Soil conservation agencies estimate the magnitude of soil erosion from measurements of the amount of removal and deposition of soil that has taken place and the size and number of gullies and sand hummocks and dunes. These measurements are used to tell how much land degradation has occurred in the past. Repeated measurements would show degradation trends.

Estimates or measurements of the amount of soil erosion on a particular piece of land can be used to estimate changes in other soil characteristics such as soil organic matter content and plant nutrient status. Soil analyses can also be made to determine the level of organic matter and plant nutrients, but they are time-consuming and require a degree of sampling and testing proficiency that is not readily available in many countries.

Additional soil characteristics have been identified by the United Nations Food and Agriculture Organization (FAO) for evaluation in its program of assessing soil degradation throughout the world (FAO, 1977). Those applicable to dryland farming are two physical properties:

sealing of the soil surface (surface crusting) and compaction of soil horizons (formation of plowpans, principally).

Wherever saline seepage is a problem, soil salinization and depth to water table would need to be assessed, also. Saline seepage on rainfed cultivated lands is usually confined to the wetter part of the semi-arid zone and is not a common problem even in that zone.

An indirect indicator of desertification is provided by crop yields. However, the interpretation is complicated by yield differences caused by drought, insects, diseases, hail, and cold weather rather than by land degradation.

IRRIGATED LANDS

Inappropriate water management is the main cause of desertification on irrigated lands. Salinization and waterlogging are the manifestations of poor water management. They outweigh by far the effect of wind and water erosion although erosion may be a problem where sandy soils or sloping land are irrigated. Even where erosion is a problem, it does not have the impact on crop yields that salinity and a high water table do. Soil compaction due to machinery undoubtedly is a common problem, as is recognized in the Central Valley of California, but little is known about its extent and its effect on crop yields. Surface soil crusting also is widespread in irrigated lands.

Assessment of irrigated land degradation is, first, a matter of measuring soil salinity throughout the root depth and determining how close the water table is to the surface. Measuring soil salinity accurately requires sending soil samples to a laboratory. Determining the depth to the water table is done by digging a hole and observing at what depth free water occurs.

Other indicators of salinity are the presence of salt crusts on the soil surface and the appearance of tip and marginal leaf burn on plants. The latter, however, is not a specific indicator of salinity, and salt crusts do not appear until the soil salinity level is relatively high. Crop yields are useful but not specific indicators of salinization and waterlogging. Plant analyses for chlorides help to determine whether soil salinity levels are high.

Sometimes soil salinity measurements need to be supplemented with analyses to determine whether there is an excessive accumulation of adsorbed sodium on the soil particles. Excess adsorbed sodium has adverse effects on soil structure and plant growth when the soil salinity is moderate or low. Tests for sodium require laboratory facilities.

As is true for grazing and rainfed cultivated lands, desertification ordinarily affects only a part of an irrigated area, in a non-uniform manner. This means that sampling sites must be chosen carefully and that broad-scale surveys - as typified by the average salinity of 100 hectare fields - will not assess conditions on the worst affected areas.

DESERTIFICATION INDICATORS

Desertification damage must be properly assessed if there is to be a rational approach made to combatting the problem. In order to make such assessments, indicators of desertification must be identified and their magnitude evaluated over a period of time. Indicators can be used to assess the current status of desertification relative to potential land productivity and to assess the current rate of desertification. Desertification risk or hazard is estimated from the presumed sensitivity of the ecological system to disturbance.

Purpose of Indicators

The kind of desertification indicators that will provide guidance to managers, planners, and policy makers is the kind that can be used to determine directly or indirectly whether a particular piece of land has been degraded or whether it presently is deteriorating, staying the same, or improving under current management practices. Change is what must be determined. The indicators should enable an observer to determine the equivalent of range condition and range condition trend for all land uses, on any property. Where such indicators can be of special utility is in assisting in the evaluation of the effectiveness of management changes by individuals as well as by the managers of development projects. At present, assessments of the impact of changed practices imposed in a development project are seldom, if ever, made a part of the project structure. Accountability cannot be required if there is no way to determine, first, whether changes are occurring and, second, whether the changes are or are not due to the introduced practices. Desertification indicators, then, are crucial to the evaluation of the effectiveness of land management practices.

CRITICAL INDICATORS

Indicators of desertification fall into three classes: biological, physical, and social. The critical ones are those few that have a high information content and come closest to meeting the criteria for superior indicators (sensitive, simple, and quantitative).

The initial effort to identify critical indicators of man-made desertification appears to have been undertaken by Berry and Ford (1977). They proposed an indicator monitoring system operating at the global, regional (international), national, and local level. Five global, five regional, and three national/local indicators were selected to embrace climate, soil, plant, animal, and human factors affecting or responding to desertification (Table 3.2). Several of the indicators actually are groupings of more specific indicators. The report outlines in considerable detail a massive program to monitor desertification.

A second attempt to identify critical indicators of desertification was initiated at a seminar held in Nairobi, Kenya immediately prior to the United Nations Conference on Desertification in 1977. The seminar was sponsored by six science associations from Africa, Asia, Europe, and North and South America; about 40 scientists participated. Results of the seminar were published in 1978 (Reining, 1978). Indicators were grouped under the headings of physical, biological/ agricultural, and social indicators (Table 3.3). The physical and biological/agricultural indicators are intended to be used to evaluate on-site (local) desertification although they could serve as the basis for assessment of land degradation for larger areas, as well.

TABLE 3.2. Critical indicators of desertification, by monitoring scale*

Monitoring scale		
Global	a.	albedo
	b.	dust storms
	c.	rainfall
	d.	soil erosion and sedimentation
	e.	salinization
Regional	a.	productivity
	b.	standing biomass
	c.	climate
	d.	nutrition
	e.	salinization
National/local	a.	productivity
	b.	human well-being
	c.	human perception

* Source: Berry and Ford, 1977.

TABLE 3.3. Critical indicators of desertification, by type*

Type	Indicators
Physical	Soil
	a. Effective soil depth
	b. Soil organic matter
	c. Soil crusts
	d. Dust, dust storms, and sandstorms
	e. Salinization and alkalization
	Water
	a. Depth to and quality of groundwater
	b. Area of standing open water
	c. Status of surface drainage systems
	Other
	Relative reflectance of land
Biological/Agricultural	Vegetation
	a. Cover
	b. Above-ground biomass
	c. Yield
	d. Key species: distribution and frequency
	Animal
	a. Key species
	b. Population of domestic animals
	c. Herd composition
	d. Production

TABLE 3.3 (continued)

Social	Land and water use
	a. Irrigation
	b. Dryland agriculture
	c. Pastoralism
	d. Mining
	e. Firewood
	f. Water
	Settlement patterns
	a. New settlement
	b. Expansion of settlement and sedentarization
	c. Diversification of settlement
	d. Settlement abandonment
	Human biological parameters
	a. Population structure and rates
	b. Measures of nutrition status
	c. Public health indices
	Social process parameters
	a. Conflict
	b. Migration
	c. Redistribution patterns
	d. Marginalization
	e. Cash versus subsistence

Kinds of Indicators

Monitoring land resources requires a determination of what is to be monitored. That, in turn, calls for the selection of indicators of the current status of the resource and an evaluation of the relation of the current condition to past conditions. Indicators may be direct (proportion of decreasers in the plant cover) or indirect (number of grazing animals) and quantitative (tons of soil eroded away) or qualitative (range condition class). Indirect indicators are undesirable because of the amount of informed or uninformed speculation that goes into an interpretation of the information collected. Qualitative indicators are, perforce, estimates, and estimates differ with the background of individuals making them. However, an experienced and knowledgeable person can provide a highly useful qualitative evaluation of the status of a direct indicator.

In the Berry and Ford (1977) report, two of the global indicators (albedo and rainfall), three of the regional indicators (productivity, climate, nutrition), and all of the national/local indicators are indirect indicators. Since satellite imagery is emphasized as the primary source of information on all the global indicators and about half of the regional indicators, the data collected would be only qualitative in character, even for the direct indicators of desertification. Scarcity of ground truth is a serious limitation to the use of satellites for data gathering. Several of the indicators (dust storms, soil erosion, standing biomass, etc.) are direct indicators only if the effect of drought, for one thing, is eliminated as a variable.

Of the critical indicators identified at the Nairobi seminar (Reining, 1978), all of the social and animal indicators, one of the vegetation indicators (yield), and half of the physical indicators are indirect connotators of man-made desertification. Again, several of the direct indicators can be strongly influenced by drought as well as by disease and insect infestations, among others.

Interpretation of Indicators

It obviously is difficult - if not impossible - to identify unambiguous direct indicators of desertification in non-irrigated land. Drought complicates the interpretation of all indicators except those of salinization and waterlogging. The complexity of separating land management effects from drought effects can be illustrated by an example from rangelands. The onslaught of a drought can cause plants to die, seedlings to fail to become established, and plant growth to stop even on the best managed ranges. After two or more consecutive years of a moderately severe or severe drought, ground cover is lessened and the soils become more susceptible to wind erosion. Since winds tend to be stronger during drought periods, dust storms normally are more frequent, more intense, and of longer duration at that time. An increase in dust storms is a reasonably reliable indication of deterioration in vegetative cover, but not necessarily an increase in man-made desertification. Isolating drought influences from management influences is not easy.

Crop and animal yield is an ambiguous indicator of desertification, not only due to the effect of weather variables but to disease and insect plagues and private or public plant and animal protection schemes. Tsetse fly infestations are responsible for keeping large parts of African savannas in good or excellent condition by excluding livestock. In that case, a low livestock yield means little desertification and the tsetse areas represent, in a sense, plant reserves waiting to be tapped. Locust swarms can devastate crop and pasture lands irrespective of their ecological condition, reducing yields to zero even in the absence of any mismanagement of the land.

It probably is safe to say that the only unambiguous indicators of desertification are an increase in the salinity level of the soil and a rise in the ground water table. Those two are indicators of salinization and waterlogging under any kind of land use, most commonly that of irrigated agriculture.

Physical and Biological Indicators

Table 3.4 represents a list of direct and indirect physical and biological indicators that appear to be the most useful and uncomplicated in an analysis of desertification. Their selection was made on the premise that quantity and quality of plant production are the most important considerations in evaluating the health of the soil and plant resource. Soil erosion is looked upon as the result of deterioration in the plant cover, since inadequate plant cover is the principal factor responsible for accelerated wind and water erosion (Talbot, 1938).

DESERTIFICATION INDICATORS 75

TABLE 3.4. Physical and biological indicators of desertification

Land use	Indicator	Type of indicator	Primary variable	Measurement procedure	Secondary variables
Irrigation	Salinization	Direct	Soil salinity	Electrical conductivity and sodium adsorption ratio of soil samples taken at various depths.	Leaf symptoms (color and tip burn), leaf chloride and boron content, crop stand, salt crusts, soil permeability. Crop yield.
	Waterlogging	Direct	Depth to water table	Depth to open water standing in hole in ground.	Plant symptoms of oxygen deficiency. Soil mottling in root zone. Crop yield.
	Soil compaction	Direct	Water intake	Infiltration rate of soil, permeability.	Restricted root development. Stunted plants.
	Surface soil crusting	Direct	Crust strength	Modulus of rupture, penetrometer.	Impaired seedling emergence. Reduced water intake.

TABLE 3.4 (continued)

Land use	Indicator	Type of Indicator	Primary variable	Measurement procedure	Secondary variables
Dryland farming	Accelerated soil erosion	Direct	Soil removal and deposition	Depth to reference subsoil horizon.	Soil color, soil reflectance, number and intensity of dust storms. Stream channel erosion.
				Number and size of rills and gullies along transect.	Stream turbidity. Soil deposition on lowlands. Crop yields.
				Number and size of hummocks and dunes along transect.	
	Nutrient deficiency	Indirect	Plant nutrient content	Analysis of plants for essential mineral nutrient content.	Crop yield. Plant symptoms of deficiency or toxicity. Spatial variation in plant growth.

TABLE 3.4 (continued)

Land use	Indicator	Type of Indicator	Primary variable	Measurement procedure	Secondary variables
	Soil compaction	Indirect	Soil nutrient content	Analysis of soils for available plant nutrient content.	Restricted root development. Stunted plants.
		Direct	Water intake	Infiltration rate of soil, bulk density.	Impaired seedling emergence. Reduced water intake.
	Surface soil crusting	Direct	Crust strength	Modulus of rupture, penetrometer.	
Pastoralism	Vegetation degradation	Direct	Plant species composition and frequency	Plant species distribution along line transects and in plots.	Animal take-off, milk production, numbers of animals (domestic and game), herd composition, soil reflectance, carbohydrate reserves in plants, and basal area of plants. Carrying capacity. Woodcutting

TABLE 3.4 (continued)

Land use	Indicator	Type of Indicator	Primary variable	Measurement procedure	Secondary variables
		Indirect	Plant biomass	Plant weight in plots.	
		Indirect	Plant vigor	Annual herbage production, seedling establishment, and amount of litter in sample plots.	
	Surface soil crusting	Direct	Crust strength	Modulus of rupture, penetrometer.	Reduced water intake. Increased runoff. Falling water table.
	Accelerated soil erosion	Direct	Soil removal and deposition	Depth to reference sub-soil horizon. Number and size of rills and gullies along transect. Number and size of hummocks and dunes along transect.	Soil color, soil reflectance, number and intensity of dust storms. Stream turbidity, stream channel erosion. Soil deposition on lowlands. Animal take-off. Carrying capacity.

TABLE 3.4 (continued)

Land use	Indicator	Type of Indicator	Primary variable	Measurement procedure	Secondary variables
Recreation	Plant destruction	Indirect	Plant biomass	Plant weight in plots.	Soil reflectance, number and area of vehicular tracks.
		Indirect	Plant vigor	Annual herbage production, seedling establishment, and amount of litter in sample plots.	
	Accelerated soil erosion	Direct	Soil removal and deposition	Depth to reference subsoil horizons.	
Mining	Accelerated soil erosion	Direct	Soil removal and deposition	Weight of material removed and deposited.	Land reflectance. Land color.
	Plant destruction	Direct	Plant biomass	Plant weight in plots.	

TABLE 3.4 (continued)

Land use	Indicator	Type of Indicator	Primary variable	Measurement procedure	Secondary variables
	Water pollution	Indirect	Water quality	Salinity, particulate matter, and heavy metals in water.	
	Air pollution	Indirect	Air quality	Amount and kind of particulate matter in air downwind.	
	Wildlife habitat destruction	Direct	Animal and bird species composition and frequency.	Observation counts in sample areas.	
	Aesthetic loss	Direct	Appearance of excavations and waste dumps.	Density, kind, and height of plant cover and amount of water surface produced.	

The indicators listed in the table are intended to provide quantitative data on soil and plant conditions at specific sites (homogeneous areas), usually of a hectare or less in size. They may be scaled up to administrative districts, nations, and regions, but the emphasis is on identifying land condition changes that are indicative of changes in the desertification status at a well-defined site. It should be obvious that one of the most difficult tasks facing the user of the table is that of deciding what site or sites to monitor. The data collected can only be good if the sampling site is representative of the area under investigation. And comparative data are most useful when the measurements are standardized and expressed in the same units (NAS--NRC, 1962).

Indirect indicators are always subject to more than one interpretation of what a change in the indicator means. They must, then, be used with caution. Our understanding of cause and effect generally is too imprecise to permit definitive conclusions to be drawn in nearly any circumstance except for the extreme condition (e.g., no soil erosion at all, or deep gullies everywhere).

The principal objective of a program to monitor changes in the various indicators is to know whether desertification is changing for the better or for the worse. A few of the indicators can be used to determine the severity of desertification (soil salinity is the best example) but most cannot. Severity must be estimated from knowledge or inference of what soil and plant conditions were prior to the time when man began to alter his environment significantly. Such conditions can sometimes be found today in protected enclosures of religious shrines and cemeteries. More commonly, they are partially reconstructed through establishment of exclosures and biosphere reserves where man and livestock influences are excluded. Exclosures established in degraded semi-arid grasslands may regenerate a reasonable facsimile of the native vegetation after a few years of favorable moisture, whereas that regeneration might take centuries to come about naturally in the more arid regions (Garner, 1950), especially if the soils are badly degraded. Soil regeneration will not occur in anyone's lifetime, so it becomes necessary to find undisturbed sites for comparison purposes.

Socioeconomic Indicators

Socioeconomic indicators are listed in Table 3.5. All of them are indirect indicators of desertification. Drought is the most common complicating factor in determining whether environmental deterioration and a reduction in human well-being is due to man-made desertification or to other forces. Great care must be taken in arriving at decisions with respect to socioeconomic indicators. This is particularly true when using satellite imagery to establish changes in land use (National Academy of Sciences, 1970).

Monitoring Schedule

Current status of desertification is determined by comparing the present land condition with what it was before man's influence was exerted or with what the

TABLE 3.5. Socioeconomic indicators of desertification

Characteristic to be measured	Primary indicators	Type of indicator	Measurement procedure	Secondary indicators
Land use	Change in land use	Indirect	Amount of cultivated land abandoned, amount of expansion of dryland farming into former pastoral lands, fallow period.	Settlement abandonment, new settlements established, increase in settlement size, per capita income, age distribution of population, social services available, migration patterns.
	Change in herd composition	Indirect	Proportion of goats and camels in herds.	
Human health	Nutritional status of children	Indirect	Weight, height, and upper arm circumference.	Infant and old age mortality. Female fertility.
Human response	Local perception of environmental change	Indirect	Personal surveys of sample population.	
Wood supply	Firewood and charcoal utilization	Indirect	Amount of firewood and charcoal used in city and village.	Price of charcoal and charcoal. Distance to go for firewood.

potential is assumed to be. Current rate of desertification depends upon monitoring current changes in land condition.

Monitoring is a repetitive undertaking. The frequency with which this is done depends upon the sensitivity of the indicators to change and the influence season of the year has on the indicator. Monitoring should be done on an annual schedule, at least, for each indicator. Some of the indicators can be monitored periodically (e.g., plant biomass), while others should be monitored episodically (e.g., dust storms). Monitoring can be done on the ground or by remote sensing. The amount of detail required determines the preferred monitoring technique. A baseline condition must be established on the ground before monitoring can begin.

Assessment and Monitoring Scales

Four general methods of measuring changes in land conditions are available: ground, low-level aerial, high-level aerial, and satellite. Costs tend to decrease as the observer gets farther from the object of observation but the accuracy also decreases. For determining the current status of desertification, an evaluation of the pre-existing land condition is essential, followed by an inventory of current conditions on the ground. Current rate of desertification is determined by monitoring the land condition or some indirect indicator of land condition, probably over a three to five year period, at the minimum. Monitoring methods are a function of the map scale. Monitoring can only be effective if the baseline condition has been adequately described.

DESERTIFICATION INDICATORS

Table 3.6 shows the suitability of ground and remote sensing methods for measuring changes in the primary variables in Table 3.4 when desertification maps at a scale of 1/20,000 are to be prepared. In every case, remote sensing must be combined with ground measurements. At smaller scales, remote sensing becomes essential – despite the sacrifice in precision – simply because the cost of ground measurements becomes excessive at map scales less than 1/100,000 and the accuracy of boundary lines tends to become less. Satellite imagery provides unparalleled advantages at map scales of 1/1,000,000 or smaller for delimiting landscape units such as ergs and chotts, where there are marked spectral contrasts between units.

Satellite imagery has captured the imagination of natural resource specialists and planners, not alone because the multicolored maps make beautiful pictures. The potential utility of satellite images is so great that the technology has generated impressive interest and support. Unfortunately, exaggerated claims of what the images can do have led to a negative response among resource managers. The great merit of satellite imagery is in the synoptic (simultaneous broad areal coverage) view it presents, in the multispectral character of the data, in the near-absence of distortion over large ground areas, and in the repetitive nature of the imaging (Johannsen and Barney, 1981). Civilian satellites suffer principally from the inability to provide needed detail for large scale maps. That deficiency is a major one. The fact that interpretation of satellite data depends on knowledge of ground conditions is frequently overlooked by

TABLE 3.6. Suitability of assessment of desertification primary variables on ground and by remote sensing for map scale 1/20,000

Land use	Primary variable	Ground	Aircraft/ground low level	Aircraft/ground high level	Satellite/ground
Irrigation	Soil salinity	+	+	+	−
	Depth to water table	+	?	?	−
	Water intake	+	+	+	−
	Crust strength	+	−	−	−
Dryland farming	Soil removal and deposition	+	?	?	−
	Plant nutrient content	+	−	−	−
	Water intake	+	+	+	−
	Crust strength	+	−	−	−
Pastoralism	Plant species composition and frequency	+	+	−	−
	Plant biomass	+	+	+	+
	Plant vigor	+	−	−	−
	Crust strength	+	−	−	−
	Soil removal and deposition	+	?	?	−
Recreation	Plant biomass	+	+	+	+
	Plant vigor	+	−	−	−
	Soil removal and deposition	+	?	?	−

TABLE 3.6 (continued)

Land use	Primary variable	Ground	Assessment method Aircraft/ground low level	Aircraft/ground high level	Satellite/ground
Mining	Soil removal and deposition	+	?	?	−
	Plant biomass	+	+	+	+
	Water quality	+	−	−	−
	Air quality	+	−	−	−
	Animal and bird species composition and frequency	+	−	−	−
	Appearance of excavations and waste dumps	+	+	−	−

* + = suitable
  ? = depends upon conditions
  − = not suitable

enthusiastic proponents of the technology. Combined with land use, geologic, soil, vegetation, and hydrologic maps, satellite sensing of the environment can be very helpful in assessing natural resources, especially in monitoring changes in surface conditions. Repetitive aerial photography would be even better at detailed scales of surveillance but the cost is usually prohibitive.

Monitoring desertification by aerial and satellite sensing in rangelands where the average annual precipitation is about 200 mm is not a simple matter (Warren and Hutchinson, 1982). Single-date reflectance data should be able to detect changes in surface soil litter (an indicator of range condition) but multiple-date observations are necessary to distinguish shrub cover from grass cover and annual plants from perennial plants. Multispectral imagery is superior to black-and-white photos in separating plant species having different phenologic characteristics.

On-the-ground measurements of desertification variables is essential to establish base conditions. Once a satisfactory data base has been constructed, monitoring certain changes, but not all, can be done remotely. Remote sensing is a technique that makes feasible repetitive surveys of large areas.

Measurement and Interpretation

Reasonably satisfactory methods are available to measure the primary variables in Table 3.4. Less satisfactory methods are available for most of the secondary variables. Qualitative rather than quantitative evaluations of the latter may be all that can be made. In all cases, on-the-ground measurements require careful sampling and are time-consuming. As a result, few assessments of desertification are based upon a thorough inventory of undisturbed and disturbed conditions. Instead, comparisons are made qualitatively - for the most part - with undisturbed conditions nearby or with reports of how conditions were prior to man's intervention. Current conditions also are assessed qualitatively and inferences are made about the severity of the degradation. For example, the amount of wind erosion that has occurred may be estimated by observing the incidence and intensity of dust storms or the appearance of small nebkas (sand accumulations) downwind of the stems of plants (Mainguet and Cossus, 1982). Rarely would detailed maps of desertification be based upon a comprehensive grid of the entire area of concern. The time and cost of doing so would be prohibitively high.

Lists of indicators of desertification and the primary and secondary variables that can be used to determine the degree of change that has occurred are useful as guides to what must be evaluated. However, collecting masses of data on the indicators may provide the analyst a greater degree of confidence but is unlikely to be necessary or cost-effective. A few well-selected samples and a careful analysis by experienced persons of interrelations among the many environmental factors involved is more valuable than reams of disconnected numbers.

Informed interpretation, then, is the key to successful evaluation of the current status, current rate, and risk of desertification. The day is still in the future when a recipe can be formulated to evaluate desertification of land units. Good progress is being made in development of models of water erosion but not of other kinds of land degradation.

Social indicators frequently are difficult to interpret. Of those, changes in land use usually are the most meaningful. Monitoring changes in land use can easily be done on the ground but it may require several years of observations to ascertain trends. In general, social indicators are of minimum value for measuring short-term changes. At the same time, it must be recognized that the effects of desertification on people is the most important consideration and, therefore, socioeconomic conditions are the measure of the true severity of desertification, in human terms.

Extension to Larger Areas

Desertification indicators identified in Tables 3.4 and 3.5 are intended for local use. They can be used also for larger areas, including global evaluations, if the means of measuring them is broadened. Thus, soil erosion remains an important indicator but the means of measuring may be changed from the precise determination of depth to an underlying layer to the calculated determination from erosion formulas (models) of the amount of erosion that is expected to occur under a given set of circumstances. Accuracy decreases, inevitably, but there is no way to carry out meaningful on-the-spot measurements everywhere in the world.

Desertification Assessment and Mapping

The United Nations Environment Program (UNEP) and the Food and Agriculture Organization of the United Nations (FAO) are engaged in a project to develop a methodology that can be used for desertification assessment and mapping. The methodology is to be applicable at the local, national, and regional scale. This means, roughly, at scales of 1/10,000, 1/100,000, 1/1,000,000, and, perhaps, 1/5,000,000. FAO published the methodology report in 1983. Assessment criteria are established for four desertification processes: vegetation degradation, water erosion, wind erosion, and salinization. Field evaluations of soil compaction are so subjective and the magnitude of the problem is so uncertain that practical criteria could not be formulated.

Soil Degradation Assessment

Soil degradation is the major desertification factor on rainfed croplands and irrigated lands. Vegetation degradation is the dominant factor on rangelands. The FAO World Assessment of Soil Degradation is intended to develop a methodology that could show the present state and the risk of soil degradation throughout the world. The principal criteria proposed for classifying land according to its present and potential soil degradation status are given in Table 3.7. Because the methodology is designed for use with FAO soil maps of the world at a scale of 1/5,000,000, it has turned out to be a risk evaluation, only. No field data are collected; all evaluations are calculated from available information on environmental factors. Present state of soil degradation cannot be assessed by using the proposed methodology. No field testing has been attempted.

TABLE 3.7. Degradation criteria proposed for FAO World Assessment of Soil Degradation project*

Degradation		
Water erosion	1)	Sheet and rill erosion
	2)	Gully erosion
	3)	Movement of mass landslides
Wind erosion	1)	Deflation
	2)	Accumulation
Excess water	1)	Shallow groundwater
	2)	Recurrent floods
Excess salt	1)	Salinization
	2)	Alkalization
Loss of chemical fertility	1)	Loss of organic matter and humus
	2)	Loss of nutritive elements and acidification
	3)	Toxicity
Physical degradation	1)	Sealing of surface (crusting) and subsoil compaction
	2)	Induration of subsoil horizons

* FAO, 1977

A world map of desertification hazards (risk, vulnerability) in the arid regions, based on potential soil degradation under current human and livestock pressures, was distributed at the United Nations Conference on Desertification (FAO/UNESCO/WMO, 1977). The criteria used for potential soil degradation were soil order, relief, climax vegetation, and climate. Soil degradation hazards - not actual degradation - were inferred from the combination of criteria that were employed. Indicators of soil changes were not utilized in delineating mapping units. Vegetation degradation was not assessed.

CONCLUSIONS

Assessment of the degree to which desertification is getting worse or is being reversed is essential for decision-makers and others who are responsible for coping with the problem of land degradation. Reliable indicators of change over relatively short periods of time are needed to determine whether current management practices are beneficial, harmful, or of a neutral character. Selecting and monitoring indicators that are sensitive to small changes in land conditions is very difficult or, perhaps, impossible with our imperfect knowledge of what to measure and how to do it. Nevertheless, an effort must be made if progress is ever to be achieved. Major changes are easy to detect.

There are many useful indicators of desertification. Their weakness lies either in the great amount of time and expense their use requires or in their inability to differentiate between the effects of weather variations and the effects of man. Only direct on-the-ground measurements of indicators are reasonably satisfactory now in evaluating change. Satellite imagery has great promise but a weak performance record, thus far, in assessing small changes. Low-level and high-level airplane photography have proved to be very useful. Cost factors militate against use of on-the-ground and airplane photography over large areas.

REFERENCES

Berry, Leonard, and Richard B. Ford. 1977. Recommendations for a system to monitor critical indicators in areas prone to desertification. The Program for International Development, Clark University, Worcester, Massachusetts. 121 p.

FAO. 1977. World Assessment of Soil Degradation - Phase I. Assessing Soil Degradation, FAO Soils Bulletin 34. Food and Agriculture Organization, Rome, Italy. pp. 29-35.

FAO. 1979. A provisional methodology for soil degradation assessment. Food and Agriculture Organization of the United Nations, Rome, Italy. 84 p.

FAO/UNESCO/WMO. 1977. World Map of Desertification. United Nations Conference on Desertification A/CONF. 74/2. Map scale: 1/25,000,000, with explanatory note.

Gardner, J. L. 1950. Effect of thirty years of protection from grazing in desert grassland. Ecology 31: 44-50.

Gay, Charles. 1965. Range Management. Why and How. New Mexico State University Cooperative Extension Service Circular 376. 32 p.

Johannsen, Chris J., and Terry W. Barney. 1981. Remote sensing applications for resource management. Journal of Soil and Water Conservation 36: 128-134.

Mainguet, Monique, and Lydie Cossus. 1982. Desertification indicators in the Sahel of Niger and Upper Volta - Methodologies and case studies. In Yusuf Elmehrik, Fakhruddin Abdulhadi Daghestani, Harvey Croze, and Wissam Al-Hashimi (editors). Environment of Monitoring for the Arab World. The Royal Scientific Society, Amman, Jordan, pp. 92-122.

Martin, S. Clark. 1975. Ecology and Management of Southwestern Semidesert Grass-shrub Ranges. Rocky Mountain Forest and Range Experiment Station, Forest Service, U.S. Department of Agriculture, Fort Collins, Colorado, Research paper RM-16. 39 p.

NAS-NRC. 1962. Basic Problems and Techniques in Range Research. National Academy of Science, National Research Council Publication No. 890, Washington, D.C. 342 p.

National Academy of Sciences. 1970. Remote Sensing, National Academy of Sciences, Washington, D.C. 424 p.

Reining, Priscilla. 1978. Handbook on Desertification Indicators. American Association for the Advancement of Science, Washington, D.C. 141 p.

Talbot, M. W. 1938. Indicators of southwestern range conditions. U.S. Department of Agriculture Farmers' Bulletin No. 1782.

Warren, Peter L., and Charles F. Hutchinson. 1983. Indicators of rangeland change and their potential for remote sensing. Office of Arid Land Studies, University of Arizona. Journal of Arid Environments (in press).

Wischmeier, Walter H., and Dwight D. Smith. 1965. Predicting rainfall-erosion losses from cropland east of the Rocky Mountains. Agriculture Handbook No. 282, U.S. Department of Agriculture, Washington, D.C. 47 p.

Woodruff, N. P., and F. H. Siddoway. 1965. A wind erosion equation. Soil Science Society of America Proceedings 29: 602-608.

4. CAUSE AND EFFECT

Desertification is caused by overgrazing, excessive woodcutting, land abuse, improper soil and water management, and land disturbance; its effects appear as reduced productivity of land, environmental degradation, impaired health, and a lowered standard of living for the local people. The problem is an old one that has become worse in the past fifty to one hundred years as land pressures have increased. Solutions are generally known but the finances and the managerial ability necessary to implement them are not always available.

The reasons why land degradation has occurred in different places varies even though the causes (overgrazing, poor drainage, etc.) may be the same. In recent years, the African continent has been the center of attention in reports on desertification. There, greatly increased populations of livestock and humans are credited with being the main source of the problem. Yet, even in Africa the reasons why degradation has occurred are not the same everywhere. And they may, in fact, be quite different than conventional wisdom would indicate.

CLIMATE CHANGE

Among the effects of desertification is the destruction of vegetative cover and an increase in spectral reflectance (albedo). One of the most striking evidences of the effect of overgrazing on albedo was seen in satellite images of the Sinai-Negev border (Otterman, 1981). Sagan et al.

(1979) suggest that anthropogenic environmental changes, including accelerated wind erosion, have been responsible for climatic changes during the past several millenia. Proof is difficult to obtain but the evidence is highly suggestive, on a regional scale if not global. Toon and Pollack (1980) concluded that soil, soot, and sulfates arising from human activity probably are warming some regions of the earth while cooling others. The complexity of the atmospheric system is too great to allow definitive conclusions to be drawn, at present.

RANGELANDS

Too many livestock concentrated on too little land area leads to overgrazing. One or two years of abuse during a drought seldom causes permanent damage to the land resource. It is the long-continued practice that is responsible for the great difficulty in restoring some lands to their former productivity and for the bleak future that many pastoralists face. In fact, restoring land to its former productivity may be impossible where severe desertification has occurred, the reason being that the soil is no longer the same as it was before desertification became serious. There is no economic way to make gullied grazing land as productive as it was before the gullies appeared.

Degraded range land is a legacy of the past in some regions of the world, the product of the present in others, and a continuation of land abuse over the centuries in still other regions. Several factors, singly or in combination, may be responsible for depauperation of range vegetation but overgrazing is the principal cause for the fact that the bulk of the world's rangelands are in poor condition and producing at less than half their potential for animal products.

Middle East

In the Middle East, the more accessible areas in the Israeli Negev Desert have been intensively grazed during times of peace over several thousands of years (Schechter and Galai, 1977) by the sheep and goats of villagers and nomads. Many nutritious and palatable plant species appear to have been completely eradicated. Removal of trees and shrubs to provide fuel and construction materials for the railroad built by the Turks in World War I was but the last instance of the deforestation that began with the destructive felling of the cedars of Lebanon, Syria, and Jordan 3,000 or more years ago. Regeneration of the woody vegetation has been hampered by accelerated water erosion and the destruction of seedlings by goats. Range deterioration was accelerated again after World War I when movement of nomads across state boundaries was discouraged. Unable to follow their customary pattern of movement, nomads were forced to graze their livestock year-around in a restricted area, with no rest period for the vegetation.

Deterioration of range lands in Iran has become rapid in the last 100 years, and particularly in the past 30 years. A study reported by Pearse (1971) estimated that the number of range animals in 1964 was twelve times the grazing capacity of the usable range. Nemati (1977) estimates that livestock numbers in the steppe zone are at least 5 times the carrying capacity. Overgrazing attendant upon a rapidly growing population of pastoralists and their herds of sheep, goats, and camels, as well as woodcutting for fuel and construction, have been responsible for much of the degradation (Department of the Environment, 1977). In addition, nationalization of range lands in 1963 and the creation of biosphere reserves may have contributed by reducing responsibilities of pastoralists for grazing land management. Aside from the loss in range productivity accompanying overgrazing, wind erosion has increased to the point where sandstorms have become a major nuisance. The effect of these problems has been to encourage the young to move to cities and to discourage investments by the pastoralists.

Niger

A case study of grazing conditions in the vicinity of Agadez, Niger, on the south side of the Sahara, enumerates an assembly of factors that accelerated desertification in recent decades in the 100 to 350 mm rainfall zone (Bernus, 1977). Among them are a breakdown in the control of the Tuareg nomadic society over what had been its grazing territory, an increase in stock numbers due to improved veterinary services and the entrance of other nomadic groups into Tuareg territory, northward expansion of dry farming into formerly good pastoral lands, population increase and consequent livestock increase among the nomadic people, and the construction of deep wells equipped with pumps.

Ovgrazing did not occur everywhere; it was severe

within 10 to 12 kilometers of major watering points but of little significance in the more remote and less accessible pastures. Since the great drought of 1969 to 1973, herds have been reconstituted and grazing pressure has risen again during the wet years that followed, in an old pattern (Killian, 1949). One notable trend underway before the drought is the growth of human and livestock population centers. That growth continues, throwing an increasingly greater burden on the environment around settlements.

Concentration of desertification around watering points and settlements produces a spotty pattern which is typical of desertification nearly everywhere. Ordinarily there is no broad advance from affected areas into unaffected areas along a line. Instead, land degradation tends to expand irregularly outward from a center point which may be a water well, a village, or a single eroded field. The spotty character of desertified land is readily seen on aerial photographs showing oil and gas wells, salinized irrigated land, watering points, and places where vehicles congregate.

Australia and the United States

Human population increases and the associated increases in livestock numbers have played a major role in desertification and certainly are responsible for a recent acceleration in the phenomenon on grazing lands. Yet, desertification affected vast areas in the United States and Australia, as well as other regions, 50 to 100 or more years ago when the human population pressure was low.

In the sparsely settled Gascoyne Basin of western Australia, sheep numbers were excessive as long ago as 1910 (Williams et al, 1977). Shipping wool to overseas markets was highly profitable then, and large flocks were an economic asset. By 1972, 15 percent of the basin was so badly degraded that continued grazing would have led to irreversible desertification. About 52 percent was partially degraded and in need of improved management, while 33 percent was still in an acceptable condition. Most of the 33 percent was hilly or stony pastures of low natural productivity. The worst affected range land, on the other hand, was the low lying and most accessible land with the highest natural productivity. It is ironic that government intervention to halt further degradation was brought about not by ranchers seeking relief but by townspeople and irrigation farmers at the mouth of the Gascoyne River who were the victims of disastrous floods attributed to excessive runoff on the degraded catchment. An economic catastrophe was required to bring about corrective measures.

Overgrazing in the western United States range lands occurred in circumstances similar to those described for Australia, with cattle the dominant type of livestock in the early years. By the 1870s and 1880s a radical and frequently permanent deterioration in vegetative cover had been brought about by the excessive numbers of cattle being fattened - at little cost - for eastern markets (Stoddart et al, 1975). Although settlement of the arid west was underway at that time, human population numbers were low. The impetus for cattle herd expansion came from

the profitability of meat production when newly constructed
railroads provided transportation connections with
population centers in the east. Overgrazing continued in
the following decades, especially at the end of a series
of wet years and the beginning of the inevitable series
of dry years that followed. By the 1920s moderate to
severe desertification was prevalent nearly everywhere and
range productivity had reached a minimum. Improvement
since then has been slow. In 1972, 72 percent of the
255,000,000 ha of rangeland surveyed was producing less
than 50 percent of its forage potential (Agricultural
Research Service, 1974).

In both Australia and the United States, soil
erosion by wind and water were serious problems on
certain landscapes but the main symptom of desertification
was the deterioration of the quantity and quality of
forage. Perennial plants were replaced by short-lived
annuals, unpalatable or unnutritious shrubs increased in
number or invaded the depleted ranges, and large barren
patches developed that shed water rapidly and worsened
the flood hazard. In neither the United States nor
Australia is sand dune encroachment on settled lands the
problem that it is in arid regions of Africa and Asia.
Gully erosion by water has had a serious impact in the
semi-arid climatic zones, especially where soils are
derived from shale formations.

Chile

Land tenure systems have had a powerful influence on
environmental stability in South America and elsewhere.

CAUSE AND EFFECT

The situation described for the Norte Chico in a case study of the Coquimbo area of Chile is reflected in most of the mining territories along the Pacific Coast. Around Coquimbo, which was first settled by Europeans in the 16th century, the large landholdings of the wealthy class have remained undivided down through the centuries or even increased in size by purchase and consolidation. Small holdings have been subdivided, as a result of land parcelling through sales and inheritance accompanying population increases, until most of them are much too small to be economically viable units (Instituto de Investigaciones Agropecuarias, 1977). In the latter case, village lands belonging to the community were surrounded by large estates. Under the circumstances, expansion of land area as the population grew was not possible, which led to a fragmentation of individual holdings.

Developmental history in the Coquimbo region exemplifies what commonly happens in mining territories. The first pre-mining settlers introduce herds of livestock in numbers that initially are small and have little adverse environmental impact. When mineral deposits are found and mining begins, roads and railroads are constructed to bring in people to staff the mines and to carry out the ore. Local demands for food increase, livestock numbers and the cultivated area expand, woodcutting to meet domestic and mining needs depletes the protective cover that trees and shrubs provide, and land degradation is accelerated.

As degradation proceeds, the livestock composition changes from cattle to sheep to goats, goats being the hardiest survivors of the selection process. At some point, the labor population stabilizes as mine employment levels off and agricultural production remains constant or declines. Out-migration takes away the young men, leaving behind women, the old, and the very young to subsist on whatever their land will produce and on money sent home by relatives working elsewhere. Stagnation prevails until another economic development starts a new cycle.

China and the Soviet Union

Case studies of desertification in western China and the Turkmen Republic of the Soviet Union further document the effects of overgrazing and woodcutting. Mobilization of sand dunes in the Maowsu sandy land of Inner Mongolia (Government of China, 1977) and in the Turfan Basin of the Xinjiang Uighur Autonomous Region (Office of Environmental Protection, 1977) is the result of land abuse. That occurred as much as 200 years ago. Protecting cultivated land of the desert oases and increasing agricultural production have required an intensive program of sand dune stabilization on the desertified perimeters of oases. Similarly, improving depleted rangelands of the steppes of northern and western China has recently received increased attention.

In Turkmenistan, desertification was a recognized problem many decades ago (Babaev and Nechaeva, 1977). The causes were the familiar ones of overgrazing and woodcutting, plus inadequate drainage in old and new irrigated oases. The spotty nature of desertification is shown once again in the much greater degradation of grazing lands in the more accessible and better watered places. Sand dune movement has accompanied the destruction of the perennial plant cover in the Kara Kum Desert, and water erosion has been a result of the same kind of destruction on the slopes of the Kopet Dagh Mountains in the south.

Recent Developments

Ancient though the overgrazing problem is, it is well to note that land degradation prior to the last several decades was not as widespread as it is now. Earlier, livestock were concentrated around villages and the few places where perennial or seasonal water was available. Nomads grazed remote areas, but only as long as surface water supplies lasted. Moreover, nomad herds were small because of the long distances between grazing sites. Two developments changed all that: 1) availability of four-wheel-drive vehicles capable of going nearly anywhere and 2) installation of deep wells equipped with powerful pumps that provided water all the year around. Those two changes, along with improved health and veterinary services, spread people, livestock, and desertification far beyond their previous confines. Railroad construction into arid regions played a similar role in spreading livestock and desertification.

Attempts to diversify the economy in pastoral areas for the purpose of reducing reliance on livestock and, thereby, easing grazing pressures, have sometimes led to a worsening of range desertification. The anomaly arises when pastoralists who earn money from cropping or other non-pastoral activities invest that money in more livestock, which defeats the purpose of the economic diversification. The reason for doing so is simple: there are no viable alternative investment opportunities (Widstrand, 1975; Pereira, 1977).

DRYLAND FARMING

Dryland farming is the kind of rainfed agriculture practiced in places where water is the principal factor limiting plant growth. The scarcity or unevenness of precipitation requires special management practices to make the most effective use of the water supplies. Among those special practices are fallowing for one or more years to store moisture in the soil and/or to make nitrogen available to the crop, wide spacing of plants in and between rows, installation of water-retaining terraces, growing drought tolerant crops, maintaining crop residue on the soil surface to reduce evaporation and erosion, and special tillage practices to conserve moisture (Cannell, 1978).

Cropping Hazards

The very nature of dryland farming makes it a hazardous practice which can only succeed if special precautions are taken. The hazards are many: a low and unreliable rainfall, dependence upon extensive rather than intensive farming, a restricted choice of crops that can be grown, hot and dry winds, soils that usually are coarse textured in the surface and highly susceptible to wind erosion, and crop yields that seldom are sufficient to justify major investment in fertilizers and pest and erosion control.

Winter and spring wheat are the dominant crops in the cooler regions; sorghum and millet are dominant in the hotter regions. Cotton, barley, and beans are other important dryland crops. Crop rotations, in the sense of a soil-building crop following a soil depleting crop, are not usually practiced, principally because suitable grasses and legumes are not available. Fertilizers are profitable only under favorable growing conditions. All in all, there is only a limited range of crop management options available to the dryland farmer (Dregne, 1978).

Erosion History

Erosion of cultivated dry lands down through the centuries has been described by numerous authors, including Lowdermilk (1948). The problem is at least two thousand years old in the Middle East, Southwest Asia, and China, where dryland agriculture has long been practiced. Pockets of land degradation were centered around villages and cities, giving a patchwork distribution similar to that noted for grazing land desertification. Erosion in dryland cropping areas was accelerated from time to time when fields were abandoned. Abandonment occurred during periods when wars erupted or droughts devastated the countryside. Water erosion probably was more significant than wind erosion in the early years of dryland farming before cultivation was extended into drier zones. Introduction of draught animals - mainly oxen - and plows made it possible for farmers to cultivate more land than they could when only hand labor was available. That contributed to erosion when it facilitated expansion of cultivation into high-risk areas. In addition, the plow had a greater adverse affect on soil stability than did hand tillage. Mechanization of agricultural production accentuated these effects.

There seems little doubt that erosion of cultivated drylands has become worse during the past 100 years. If there has been a decrease in the rate of soil degradation in some places during recent decades, it is due primarily to a delayed response to a problem that already was serious. Such has been the case in Australia, the United States, China, Tunisia, the Soviet Union, and elsewhere. Conservation practices never accompany the opening of new land to cultivation; they always come afterward, if ever.

Weather Variation

Cycles of wet and dry periods pose problems to persons and governments attempting to establish farming communities in semi-arid regions. During wet periods cultivation is extended into areas previously considered to be too dry for successful farming. Frequently, if expansion occurs early in the wet cycle, the gamble turns out to be profitable and cropping is extended even further in following years. Ultimately, the weather turns dry again and crop failures become widespread. When that happens, some or all of the new land is abandoned, with severe social and economic impacts. Then, in time, the rains return and the boom-and-bust cycle begins again (Hewes, 1975).

Dryland farmers tend to be overly optimistic about what the weather will be next year. Although their optimism may be essential if they are to continue farming, it gives them a distorted view of rainfall probabilities (Miewald, 1978). They are much more likely to believe that above-average rainfall occurs in 60 percent of the years than to accept the reality that the opposite is true.

Causes of Land Degradation

Overextension of cultivation into marginal climatic areas is not the only cause of land degradation. Cultivation of sloping lands and sandy, loessial, and shale soils which cannot possibly be kept from eroding once their native cover has been disturbed is another important cause. Failure to use known techniques of soil and water conservation such as strip cropping, terracing, and maintenance of crop residue on the soil surface contributes significantly to the erosion problem. A special kind of land degradation in Australian, United States, and Canadian drylands is caused by saline seeps which owe their origin to the replacement of perennial deep-rooted vegetation with shallow-rooted annual plants (Vander Pluym, 1978).

In general it probably is safe to say that there are few dryland farming areas in the world that have not experienced at least moderate desertification. Water erosion is the dominant problem in Mediterranean climates - where most of the precipitation falls during the cool season - and in summer monsoon climates. Wind erosion is the major hazard on sandy soils and on medium textured soils where fallowing is practiced. The principal exception to the foregoing generality is the loessial plateau of North China. That situation is unique because the thick loessial deposits stand high above the rivers that drain the region. Any steam channel in the unstable aeolian silt soon becomes deeply incised. Subsequent undercutting of the near-vertical stream banks causes sloughing of large sections of the banks. The tremendous silt load of the Yellow River is largely the product of water erosion in the loess plateau (Tieh, 1941), which is extremely difficult to control.

IRRIGATION

Salinization and waterlogging have been the curses of irrigated agriculture down through the years. Excess application of water by surface irrigation methods has been the primary cause of rising groundwater tables and the subsequent accumulation of soluble salts in the upper part of the soil profile. There are four main reasons why excess water is applied: 1) the belief that if a little water is good, more is better, 2) inadequate levelling of land, which makes efficient use of water impossible, 3) the low cost of water in most irrigation districts, and 4) the scheduling of deliveries of water on a rotation system rather than on demand. The latter leads to problems because it is conducive to irrigating fields every time water is available, whether or not the crops need it. Two additional major causes of waterlogging and salinization are seepage from unlined canals and ditches and the absence of a drainage system.

CAUSE AND EFFECT

Irrigation is commonly looked upon as the panacea for all the ills of arid land development. If only water were available in unlimited quantities, all would be well - or so it is said. In the real world, such is rarely the case for any extended period of time. And the bigger the irrigation project, the more certain it seems to be that trouble will appear, sooner or later. One reason is that planners and builders of irrigation projects seldom concern themselves with on-the-farm water management. Conventionally, dams are constructed to store as much water as possible, canals are laid out to deliver the water to various points within the area to be irrigated, and the farmer (often without irrigation experience) is left to his own devices to build field laterals, level the land, plant his crops, and decide how to irrigate his land. No district-wide provision is made for the drainage problems that will almost certainly arise in the future. With that limited approach, it is no wonder that planners are surprised when the reservoir begins to fill with silt, irrigation canals leak, crop yields do not meet expectations, and waterlogging and salinity threaten the economic viability of the project. And yet, despite the vast amount of experience with irrigation agriculture, governments continue to act as though merely building a big enough dam assures the success of an irrigation project.

Salinization and Waterlogging

Salinization of soils can occur without waterlogging, and waterlogging does not necessarily lead to soil salinization. By and large, however, waterlogging is the principal cause of salinization in irrigated areas. Using saline water for irrigation or irrigating slowly permeable soils are secondary causes of salinization. Most large rivers which serve as a source of irrigation water, such as the Nile, Indus, Yellow, Syr Darya, and Columbia, are low in salts. Salinity problems associated with use of those waters are the result of evaporation of groundwater or of the dissolving of salts originally present in soil strata. A few streams, of which the Pecos River in Texas and the Rio Salado in Argentina are examples, are highly saline during periods of low flow. Wells represent the water source that is most likely to be excessively saline.

Waterlogging is a problem all by itself because the great majority of crop plants are unable to survive if their roots are under water. Rice is the main exception. High water tables effectively reduce the root zone of plants to the depth of soil between the water table (free water) and the soil surface. If the water table is at the surface, plants drown due to the lack of oxygen (van't Woudt and Hagan, 1957).

The interrelation of salinization and waterlogging can be illustrated by an example from Turkmenistan in the Soviet Union. There, completion of a vast network of irrigation canals fed by the gigantic Kara Kum Canal has brought dual problems of salt and high water tables over many thousands of hectares.

Moderate to high levels of soil salinity affect plant growth adversely by making water intake by roots more difficult and by injuring leaf tissues (U.S. Salinity

Laboratory Staff, 1954). Plants vary greatly in their ability to tolerate salts. Barley and sugar beets are among the most tolerant and beans among the least tolerant (Ayers and Westcot, 1976).

Iraq

Perhaps the oldest documented incidence of waterlogging and salinization of irrigated land is that in the Mesopotamian Plain created by the Tigris and Euphrates rivers in Iraq. The plain is especially susceptible to waterlogging and salinization because the land is flat, evaporation rates are high, and the soils are slowly permeable silts and clays. The earliest phase of serious salinization occurred between 2400 and 1700 B.C., and that one has been followed at intervals by others up to the present time (Jacobsen and Adams, 1958). In 1950 it was estimated that approximately 60 percent of all the agricultural land in Iraq was seriously affected by salinity, with about 20 to 30 percent of that land abandoned and more being lost each year (Dougrameji and Clor, 1977).

The first attempt in modern times to establish an irrigation project that would intensify crop production on the lower Mesopotamian alluvium took place in 1953. The Greater Mussayeb Project was intended to be a large scale field demonstration of how to reverse desertification while at the same time effecting land distribution and organizing work on a cooperative basis (Dougrameji and Clor, 1977). As it turned out, the project in its early years became a demonstration of the problems that had plagued previous irrigation developments. It was not until 1968 that resources were made available to accomplish what the project had set out to do. Meanwhile, valuable - but expensive - lessons were learned. The case study chronicles, in a most illuminating way, what should and should not be done in executing an irrigation project.

Considering the fact that all the soils of the project area had been classified originally as highly saline, with electrical conductivies of the saturation extract exceeding 14 millimhos/cm, it is no wonder that it was difficult to achieve the desired results. Due to a variety of reasons, the groundwater table rose, salinity levels increased, and large areas became unproductive. Moreover, silt deposits in the canals and ditches caused water shortages in the system. Many farmers abandoned the area. All in all, the litany of problems is reminiscent of what had been happening for centuries.

In a notable departure from the usual procedure, planners of the Greater Mussayeb Project has done one thing that had never been done before in an Iraqi irrigation project: they had arranged to construct a drainage system at the same time irrigation canals were dug. However, only the main drains were built; field drains were the responsibility of the farmers, and none were installed. Failure to keep the major drains clean further exacerbated the problem. Consequently, water-logging and salinization occurred even though tests showed that the soils were sufficiently permeable to be leached effectively if water was managed properly and drainage provided.

The case study enumerates a number of reasons for

the unhappy developments during the first 12 years of the project. Those reasons include the canal and drain maintenance problems associated with the lack of needed authority, funds, equipment, and staff to cope with them; the fact that one-third of the settlers had had no previous agricultural experience; illiteracy was high, few extension services existed; the agricultural cooperatives could not meet the requirements for credit, purchasing, and marketing facilities; land tenure conditions were unsatisfactory; veterinary, health, sanitation, and education services were inadequate or non-existent; and housing design was unacceptable.

A rehabilitation scheme was initiated in 1965 and intensified after 1968 when funding was increased and managerial control strengthened. Irrigation and drainage canals were cleaned, soils were surveyed and classified, consumptive use of crops calculated or determined, climatological data were collected, an improved drainage system was designed, research and demonstration programs were undertaken, and restrictions on land ownership were tightened. As a result, considerable progress has been made in raising the standard of living for part of the population but difficulties continue to hamper rehabilitation. What needs to be done is known, but doing it is another matter when demands on national resources are heavy. The authors of the case study report emphasized the need for a pilot project to analyze and resolve problems on a small scale before beginning large scale development. There are no technical obstacles to development but social problems are not so amenable to solution.

Pakistan

The conditions under which irrigation agriculture is practiced in the Indus Plain of Pakistan are similar to those of Iraq, along with one important difference: soils of the Indus Plain are much more permeable than those of the Mesopotamian Plain. As a consequence, control of waterlogging and salinization is much easier.

A case study of the Mona Reclamation Experimental Project details events related to irrigation development in the Indus Plain, the soil and water problems that arose, and the measures taken to solve those problems (Irrigation, Drainage, and Flood Control Research Council, 1977).

Canal irrigation began in 1901. A surface drainage system was begun in the 1930s to correct waterlogging and salinization problems that were affecting significant areas. By the 1960s, 90 percent of the irrigated land suffered from high water tables and about 20 percent from excessive soil salinity. Leaking canals, inadequate land levelling, poor water management, and the absence of an effective drainage system were responsible for the difficulties.

Reclamation of waterlogged and salinized irrigated land in Pakistan was intensified in the 1960s when a number of Salinity Control and Reclamation Projects (SCARP) were established. The Mona area is a part of one of those projects. The initial step in reclamation was to install tubewells which served to lower water tables and - in most cases - to provide water of satisfactory quality for irrigation. Later, research was undertaken on a number of agronomic, water management, and economic concerns. By 1976, groundwater pumping had lowered water tables to a safe depth. Reclamation practices had been tested and soil reclamation was proceeding slowly. Several questions remain to be answered about on-farm water management, cropping practices, and socioeconomic matters but the technology for controlling groundwater tables is now available.

The varied problems encountered in setting up and carrying out rehabilitation of the Mona Project, an area of about 45,000 hectares, persuaded the authors of the case study report that a pilot program should have been used to test ideas and practices. As it was, research accompanied, rather than preceded, the introduction of reclamation practices. Consequently, answers to questions did not come in time to avoid severe disappointments among farmers and the inefficient use of resources.

Egypt

Irrigation has been carried on successfully in the Nile Valley of Egypt for thousands of years. Prior to the construction of the High Aswan Dam in the 1960s, part of the cropped land received water only during the annual floods and part received flood water plus water diverted from the river by dams (barrages) during the remainder of the year. Salinity problems had arisen in the Nile delta before the High Dam was built but waterlogging was not common. The annual flood flushed accumulated salts from the soil and helped control salinization.

One of the unforeseen consequences of the controlled perennial irrigation that the High Dam made possible was the development of a serious waterlogging problem. In 1976 it was estimated that on 80 percent of the valley and delta lands crop yields were slightly to severely reduced by waterlogging and the associated salinity (U.S. Department of Agriculture, 1976). Salinity and waterlogging have already caused as much as 20,000 hectares of newly irrigated land west of the delta to be abandoned after only a few years of cropping.

On the old lands, waterlogging was the result of overirrigation - made possible by a year-around water supply - and inadequate drainage. In the new lands, the dramatic and unexpected rise of water tables was due primarily to canal seepage, excessive percolation of irrigation water through the coarse textured soils, and the absence of a drainage system. Remedial drainage systems are being installed, at a very high cost, in both the old and new lands.

Egyptian experience with irrigation illustrates a truism that often is overlooked: the more one tries to control nature, the more difficult it is to maintain control. Uncontrolled floods wreaked havoc on Egypt periodically, but the low intensity cropping practice that it permitted could go on indefinitely with little technological investment. Once the floods were controlled and a perennial water supply assured by construction of the High Dam, a number of problems arose, of which waterlogging and salinity are only two. Control of the river was essential since there was no possibility of increasing crop production to the levels needed unless flood water could be stored for later release. And so a vicious circle develops: increasing yields demands new technology which creates new problems which require more new technology.

MINING

Land disturbance is an inevitable consequence of mining, whether it be for gas, oil, or minerals. The vegetative destruction associated with digging a shaft or a huge pit is only one part of the degradation accompanying mining. In addition to the mine opening itself, there may be piles of waste rock (from mineral separation) nearby, access roads and railroads, ore treatment plants, pipelines, and a community for families and supporting services. Each of these causes land disturbance. In all too many cases, the disturbed land remains barren indefinitely because laws do not require rehabilitation of the land or because the expense of rehabilitation is too great.

Mining has been practiced in the arid regions since time immemorial. Even so, the amount of land affected was almost infinitesimal, on a world scale, until the twentieth century. With the advent of machinery capable of excavating huge amounts of earth, and the increasing demand for metals and coal, all that is changing. Earlier in the century, open-pit copper and iron mines produced the greatest alteration of surface features; now, surface mining of coal is changing the face of hundreds of thousands of hectares of land. Surface mining has a great potential impact. The arid regions of Chile for example, contain immense deposits of copper, nitrates, coal, and iron that are at depths suitable for surface mining. In the rainless Atacama Desert, surface mining causes few changes in the environment, other than aesthetic degradation. In the less arid regions, however, the damage will be aesthetic as well as biological and physical. And the damage may well be irreversible.

The problems associated with revegetation of mined land in the arid regions can be illustrated by an analysis of the environmental effects of surface coal mining in the United States (National Academy of Sciences, 1974). Waste earth materials (spoils), which are piled up near the open pit of the mine, are composed primarily of infertile subsoils and broken rock. The original topsoil may or may not be set aside in a special operation that permits replacement of the topsoil during the mine rehabilitation process. Spoil banks have the conformation of ridges or small hills, with steep slopes.

Spoil material has a number of limitations that make it difficult to establish a vegetative cover. These limitations include such things as infertility, looseness, steep side slopes and excessive salinity. The number one limitation, however, is usually the dryness of the climate. Under the best of edaphic conditions, revegetating barren soils is difficult in places where the average annual precipitation is 150 mm or less; it may be impossible on mine spoil banks unless very expensive treatment is used, including the supplying of irrigation water (Wali, 1975). Natural revegetation is sure to be slow, requiring several decades in the semi-arid regions and several centuries in the drier regions. In the absence of vegetative cover, spoil banks are readily susceptible to water and wind erosion.

While establishing a vegetative cover on spoils is a major challenge to surface mining operations, repairing damage done in the vicinity of oil and gas wells and shaft mines, along roads, and around construction sites and communities also is difficult. In most cases, nothing is done about alleviating the environmental impact of these activities. Occasionally, the discomfort caused by sand and dust storms sweeping through communities leads to efforts to reduce the wind erosion hazard.

Shaft mines usually have one requirement that has a devastating effect on the tree population: wood for supporting roofs of tunnels. Freer (1978) says that all available timber for 40 kilometers around the mines at Cobar in Australia was cut down 50 years ago. Similar conditions existed wherever shafts were dug, mine superstructures built, or ores smelted. Improved mining technology and the availability of coal and petroleum halted wood exploitation, but only after extensive and long-lasting damage had been done.

Surface mining of coal will disturb only a small part of the arid lands of the United States. Projections to the year 2000 indicate that the disturbed land will amount to about 80,000 hectares (National Academy of Sciences, 1974), which is approximately 0.03 percent of U.S. arid lands. For persons affected by the mining operation, however, the fact that the disturbed area is small on a national scale is of little import. Surface coal mining undoubtedly will disturb much more land than the combination of surface iron, copper, and salt mining, together.

Surface mining operations have a number of environmental effects that are almost always undesirable while mining is in progress. The hole, itself, and the barren spoil banks around it are unattractive; wind erosion of spoils pollutes the air downwind and water erosion pollutes the land and streams; groundwater supplies may be interrupted and contaminated; the surface hydrology of the area is altered; plant and animal life is destroyed; and the soil mass has been changed permanently, for all practical purposes.

Much of the damage can be repaired - if cost is no limitation - but the surface and subsurface hydrology and the soil cannot be restored. If the surface depressions can be transformed into lakes that meet a need, the change in the surface hydrology can become an asset. That is not likely to be the case with either the subsurface hydrology or the soil. It becomes necessary, then, to think of land treatment in terms of rehabilitation rather than of restoration. Restoration means that the original conditions, in all respects, will prevail again. That is impossible. Rehabilitation, as used here, means that the land will be treated in such a manner that the final condition is acceptable to decision-makers, whoever they may be. That also may be impossible, in practice if not in theory. Rehabilitation differs somewhat from reclamation. The latter means - in context of land - to turn waste or low productivity land into more productive land. Rehabilitation may or may not do that.

RECREATION

Man has long been attracted to the arid regions by

the mild weather and abundant sunshine throughout much of
the year. To some people, the prime attraction may be the
oftentimes spectacular scenery, the isolation and quiet
of the desert, or the unique character of the plants and
animals. But to many of those who are escaping from the
confinement of urban life - inside or outside the desert -
the open spaces have their allure in the opportunities
they offer to unleash emotions and energies. And the last
decade has seen more and more people expending those
energies in driving off-road vehicles (four-wheel drive
automobiles, motorcycles, dune buggies, etc.) over rough
landscapes that previously were inaccessible to motorized
vehicles.

Recreational uses of the arid regions do not have to
be destructive. Hunting for attractive rocks, watching
birds and animals explore their environment, observing
colorful sunrises and sunsets, camping and hiking,
sightseeing, or just doing nothing in quiet surroundings
usually has little adverse impact on the environment.
Most visitors engage in this type of activity. Even
those basically innocuous pursuits can, however, be
damaging if too many people indulge in them at the same
place or if the visitors' conduct is irresponsible.

Off-Road Vehicles

By far the most destructive activities of the
recreational users of desert areas are associated with
off-road vehicles. These vehicles, capable as they are
of climbing hills and dunes, crossing arroyos, and riding
roughshod over shrubs, have a tremendous potential for

damage. In the desert of southeastern California, off-road vehicle environmental damage has reached the stage where controls have been instituted in an effort to prevent continuing destruction of the natural condition (Environmental Defense Fund, 1974).

The California desert is particularly susceptible to off-road vehicle damage because it is within easy driving distance of heavily populated Los Angeles. Ten million people can look to the six million hectares of desert - most of it federally owned land - as a playground and escape from the city (U.S. Bureau of Land Management, 1968). The comparatively few who choose to take their four-wheel-drive vehicles, motorcycles, and dune buggies to the desert for recreation have an impact out of all proportion to their numbers.

Vehicles roaming across the deserts are a source of both irritation and damage. Not only are they excessively noisy, they are also responsible for harrassment and death of animals, injury and destruction of vegetation, accelerated water and wind erosion, damage to archaeological sites, and other abuses. Their use had increased dramatically in the past decade and is expected to continue to increase in the future. Much of the damage they cause is essentially irreversible.

Off-road vehicle damage to arid lands is not confined to the California desert. Other regions have the same problem, but to a lesser degree. Not many deserts will find 2,500 motorcycles lined up in preparation for a race across 150 kilometers of trackless land, as has occurred in California. Australia is unfortunate enough to experience similar pressures (Robinson, 1978), and more countries will in the future.

Sightseeing

One of the more unusual instances of desertification has been caused by tourists seeking to admire sunsets and to photograph and climb Ayers Rock in central Australia. Sunset watching and photography would not seem to have much of a destructive potential. However, when tens of thousands of people choose one place to engage in either activity, the result is environmental damage.

The attraction of Ayers Rock arises from the scenic character of the Rock and the spectacular sunsets that can be seen from a neighboring sand dune. At sunset time, as many as a thousand people may congregate on the dune to enjoy the changing colors of the sky. The result of tourist traffic to Ayers Rock and Sunset Strip Dune is increased water and wind erosion. The dune has lost its vegetative cover and now is loose and mobile; the tracks and road around Ayers Rock have altered water runoff patterns, caused the death of trees, and accelerated gully erosion; and litter discarded by visitors degrades the visual appeal of the arid landscape (Freer, 1978).

COMMUNITIES

Towns and villages may have favorable or unfavorable effects on surrounding lands. The effects usually are unfavorable, but the rare community that provides a green belt of trees and shrubbery around it will have brought about improved conditions.

The basic environmental problem associated with communities of any size is that they are placed where people congregate, permanently or intermittently. And the greater the number of people, the greater the potential for damage. A livestock marketing center of a few hundred people may be inundated with thousands of sheep, goats, and cattle for several days each year. During those days, trampling and heavy grazing will denude the adjoining countryside of all vegetation and assure that no possibility of regeneration occurs before the next onslaught. The circle of land degradation that appears around a marketing center is typical of desertification patterns: a central area of severe desertification surrounded by less severe degradation, with erosion-prone trails and roads radiating out from the center. The spreading pattern of desertification occurs around villages, towns, watering points, camp sites, recreational areas, and mines, wherever people or livestock come together.

Fuel scarcity is a major factor leading to desertification around communities. Woodcutting is a never-ending business in fuel-short developing countries. It is commonplace to see a steady stream of wood-laden donkeys and camels entering Niamey in Niger and Ouagadougou in Upper Volta every day of the year. There appears to be no end to the relentless pressure on the declining wood resources.

Eckholm (1979) has noted the magnitude of firewood consumption: about half of the wood cut throughout the world each year is burned as fuel, principally by the billion or more people who rely on firewood for cooking and heating. Per capita consumption is estimated to come to between one-fifth of a ton in wood-short countries such as India to one ton in the humid tropics. One example of the degree to which woodcutting has eliminated trees and shrubs around cities is cited by Delwaulle and Roederer (1973). They speak of a tendency toward total desertification within a radius of 40 to 50 kilometers of Niamey due to fuelwood cutting. Annual wood consumption in Niamey in 1972 is said to have been about 20,000 metric tons, practically all of it brought in on camels and donkeys and sold for increasingly higher prices as supplies become scarcer.

CONCLUSIONS

The causes of desertification vary from place to place and from time to time. In grazing lands, the obvious ultimate cause of land degradation is overgrazing, if one ignores the contribution made by wood cutting. Causes for desertification of dry farming and irrigated lands are more complex although they can be collectively assigned to improper soil and water management. Mining effects are largely due to failure to rehabilitate waste deposits. Recreational abuses stem from congregation of large numbers of people and their vehicles in limited areas and a cavalier attitude toward natural resources.

Reasons for Desertification

Environmental pressures due to increasing human and animal populations certainly played a major role in bringing about desertification in recent decades in the developing countries. Yet, it is well to note that land degradation is not a new phenomenon. Erosion, salinization, and waterlogging were problems a hundred or more years ago when populations were only a fraction of what they are now. Both rich and poor countries have experienced severe desertification on at least some part of their national territory. Desertification continues to grow worse in most - but not all - developing countries. In the developed countries, the situation varies from place to place: some lands are being improved, some are undergoing further deterioration, and others have reached a state of equilibrium, usually at a low level of productivity.

If excessive human population is not the sole or - in many cases - the principal reason for desertification, what are the other reasons? There are many. One is the exploitative philosophy that sees land as a limitless resource: if one piece of land is exhausted, simply move on to better land somewhere else. Another, which is linked with population, is the social custom of using land communally. Communal use can be effective in protecting the resource if there is strong leadership and community participation, but it can lead to land abuse if the responsibility and authority to manage the resource are diffused. Still other reasons include land tenure systems that fragmentize holdings, gross over-estimation of carrying capacities, lack of technical advice on how to control desertification, absence of inducements to conserve land, high prices for grain and meat, inadequate credit arrangements for improvement of operations, and weakness or non-existence of laws and regulations governing land abuse. An underlying reason for pastoralists allowing overgrazing may be that their primary concern - in traditional and modern societies - is for the health of their animals rather than for the health of the vegetation (Department of the Environment, 1977; Pederson, 1980).

Status of Desertification

It appears, in general, that desertification of much of the arid grazing lands in the Middle East, central and southwest Asia, North Africa, the United States, Mexico, South America, and Australia reached the severe stage during or before the 19th century. Since then, conditions have continued to become worse, especially in what had formerly been relatively inaccessible or waterless regions. In Africa south of the Sahara, localized desertification was apparent toward the end of the 19th century but widespread deterioration did not become serious until the 20th century. The current condition is one of continued steady land degradation throughout most of Africa except where tsetse fly infestations exclude livestock. Once the tsetse fly is eradicated or controlled, rapid deterioration of those lands can be expected.

Dryland farming areas are experiencing continued desertification virtually everywhere. Few places show the monstrous gullying of the loessial highlands of China but water and wind erosion are ubiquitous on cultivated lands. There is an ebb and flow of desertification on the arid fringe of the dry farming areas as wet and dry cycles occur and as grain prices rise and fall. When prices and/or the weather are favorable, cultivation expands on to marginal lands where drought and erosion hazards are high. During a wet period, wind erosion may be minimal although water erosion could be considerable. When the inevitable dry period occurs, crops are poor or land is abandoned, and wind and water erosion are accelerated. Year in and year out, there is a gradual deterioration, rapid in some years, slow in others.

Salinization and waterlogging may show considerable variation from year to year because the technology and management practices for controlling them are well known. Whether or not reclamation principles are applied is a function of the funds, equipment, and advisory personnel available and the motivation of the farmer. Soil conditions, on the average, are remaining unchanged or deteriorating in most irrigated areas. They are deteriorating in eastern Australia, in the Imperial and Mexicali valleys of the United States and Mexico, in Egypt, along the Kara Kum Canal in the Soviet Union, and in Rajasthan State in India. Little or no change can be detected in most other areas although some improvement is underway in Pakistan and in limited areas elsewhere.

Desertification due to mining and recreation is increasing wherever those two land uses are practiced. Regulation of those activities frequently is either not attempted or is difficult to enforce. Neither affects large land areas but both may have locally serious environmental effects.

REFERENCES

Agricultural Research Service. 1974. Review of ARS Research Program on Western Range Ecosystems. Technical Report. Agricultural Research Service, U.S. Department of Agriculture, Washington, D.C. 108 p.

Ayers, R. S., and D. W. Westcot. 1976. Water Quality for Agriculture. Food and Agriculture Organization Irrigation and Drainage Paper 29, Rome, Italy. 97 p.

Babaev, A. G., and N. T. Nechaeva (editors). 1977. U.S.S.R. Integrated Desert Development and Desertification Control in the Turkmenian SSR. An Associated Case Study. United Nations Conference on Desertification, A/CONF. 74/22. 49 p.

Bernus, Edmond. 1977. Case Study on Desertification. The Eghazar and Azawak Region. Niger. United Nations Conference on Desertification, A/CONF. 74/14. 111 p.

Cannell, Glen H. (editor). 1978. Proceedings of an International Symposium on Rainfed Agriculture in Semi-Arid Regions. University of California, Riverside. 703 p.

Delwaulle, J. C., and Y. Roederer. 1973. Le bois de feu á Niamey. Bois et Forêts des Tropiques No. 152, pp. 55-60.

Department of the Environment, 1977. Case Study on Desertification. Iran: Turan. Department of the Environment, Government of Iran. 97 p.

Dougrameji, J. S., and M. A. Clor. 1977. Case Study on Desertification. Greater Mussayeb Project. Iraq. United Nations Conference on Desertification, A/CONF. 74/10. 102 p.

Dregne, H. E. 1978. The effect of desertification on crop production in semi-arid regions. In Glen H. Cannell (editor), Proceedings of an International

Symposium on Rainfed Agriculture in Semi-Arid Regions, University of California, Riverside, pp. 113-127.

Eckholm, Erik. 1979. Planting for the Future: Forestry for Human Needs. Worldwatch Paper 26, Worldwatch Institute, Washington, D.C. 64 p.

Environmental Defense Fund. 1974. ORVs and the California Desert: A Case Study. The ORV Monitor, June, 1974, pp. 3-6.

Freer, Peter. 1978. Non-agricultural land use and desertification. Search 9: 276-280.

Government of China. 1977. China. Control the Desert and Create Pastures. An Associated Case Study. United Nations Conference on Desertification, A/CONF. 74/17. 27 p.

Hewes, Leslie. 1975. The Great Plains one hundred years after Major John Wesley Powell. In Brian W. Blouet and Merlin P. Lawson (editors), Images of the Plains, University of Nebraska Press, Lincoln, pp. 203-214.

Instituto de Investigaciones Agropecuarias. 1977. Case Study on Desertification. Region of Combarbala, Chile. United Nations Conference on Desertification, A/CONF. 74/9. 162 p.

Irrigation, Drainage and Flood Control Research Council. 1977. Case Study on Desertification. Mona Reclamation Experimental Project. Pakistan. United Nations Conference on Desertification, A/CONF. 74/13. 120 p.

Jacobsen, Thorkild, and Robert M. Adams. 1958. Salt and silt in ancient Mesopotamian agriculture. Science 128: 1252-1258.

Killian, C. 1949. The degradation and protection of the soils in the pastures of the Algerian steppes. Proceedings of the First Commonwealth Conference on Tropical and Sub-Tropical Soils, 1948. Commonwealth Bureau of Soil Science, Technical Communication No. 46, pp. 192-195.

Lowdermilk, W. C. 1948. Conquest of the Land Through
    Seven Thousand Years. Soil Conservation Service,
    U.S. Department of Agriculture, MP-32. 33 p.

Miewald, Robert D. 1978. Social and political impacts
    of drought. In Norman J. Rosenberg (editor),
    North American Droughts, Westview Press, Boulder,
    Colorado, pp. 79-101.

National Academy of Sciences. 1974. Rehabilitation
    Potential of Western Coal Lands. National Academy
    of Sciences, Washington, D.C. 198 p.

Nemati, Nasser. 1977. Range rehabilitation problems of
    the steppic zone of Iran. Journal of Range
    Management 30: 339-342.

Office of Environmental Protection. 1977. China.
    Tame the Wind, Harness the Sand, and Transform the
    Gobi. An Associated Case Study. United Nations
    Conference on Desertification, A/CONF. 74/16. 21 p.

Otterman, J. 1981. Satellite and field studies of man's
    impact on the surface in arid regions. Tellus
    33: 68-77.

Pearse, C. K. 1971. Grazing in the Middle East-Past,
    Present, and Future. Journal of Range Management
    24: 13-16.

Pederson, Oscar. 1980. Caring for Montana's land and
    water resources. Journal of Soil and Water
    Conservation 35: 196.

Pereira, H. C. 1977. Land-use in semi-arid southern
    Africa. Philosophical Transactions of the Royal
    Society of London, B. 278: 555-563.

Robinson, C. S. 1978. Soil conservation in the Northern
    Territory. Journal of the Soil Conservation Service
    of N.S.W. 34: 101-105.

Sagan, Carl, Owen B. Toon, and James B. Pollack. 1979.
    Anthropogenic albedo changes and the earth's
    climate. Science 206: 1363-1368.

Schechter, Yoel, and Chaya Galai (editors). 1977. The Negev: A Desert Reclaimed. An Associated Case Study. United Nations Conference on Desertification, A/CONF. 74/20. 110 p.

Stoddart, Laurence A., Arthur D. Smith, and Thadis W. Box. 1975. Range Management. Third Edition. McGraw-Hill Book Company, New York, N.Y. 532 p.

Tieh, T. Min. 1941. Soil erosion in China. Geographical Review 31: 570-590.

Toon, Owen B., and James B. Pollack. 1980. Atmospheric aerosols and climate. American Scientist 68: 268-278.

U.S. Department of Agriculture, 1976. Egypt. Major Constraints to Increasing Agricultural Productivity. U.S. Department of Agriculture Foreign Agricultural Economic Report No. 120. 193 p.

U.S. Salinity Laboratory Staff. 1954. Diagnosis and Improvement of Saline and Alkali Soils. U.S. Department of Agriculture Agriculture Handbook No. 60, Washington, D.C. 160 p.

Vander Pluym, Henry (editor). 1978. Dryland-Saline-Seep Control. Proceedings of the Subcommission on Salt-Affected Soils. 11th International Soil Science Society Congress, Edmonton, Canada. 315 p.

Van't Woudt, Bessel D., and Robert M. Hagan. 1957. Crop responses at excessively high soil moisture levels. In James N. Luthin (editor), Drainage of Agricultural Lands, American Society of Agronomy, Madison, Wisconsin, pp. 514-611.

Wali, Mohan K. (editor). 1975. Practices and Problems of Land Reclamation in Western North America. The University of North Dakota Press, Grand Forks. 196 p.

Widstrand, Carl Gosta. 1975. The rational of nomad economy. Ambio IV: 146-153.

Williams, O. B., H. Suijdendorp, and D. G. Wilcox. 1977. Australia. Gascoyne Basin. An Associated Case Study. United Nations Conference on Desertification, A/CONF. 74/15. 177 p.

Williams, R. E. B., B. W. Allred, Reginald N. Denio, and
    Harold A. Paulsen, Jr. 1968. Conservation,
    development, and use of the world's rangelands.
    Journal of Range Management 21: 355-360.

5. PREVENTION AND REVERSAL OF DESERTIFICATION

The old adage that an ounce of prevention is worth a pound of cure applies as well to desertification as it does to human ills. Unfortunately, however, the attention of people and governments is more easily directed toward coping with disasters than to preventing them. Founding of the U.S. Soil Conservation Service in the 1930s was the result of well-publicized reports on the disastrous effects excessive water erosion was having on the land and the lives of people in the southeastern United States. Warnings issued decades before by knowledgeable people about the impending crisis were ignored, just as they have been about petroleum supplies, endangered species, environmental pollution, and other matters.

Desertification is now so widespread in the arid regions that a massive effort is needed to improve degraded land. Prevention of further deterioration should always be the first consideration, whether the land is in good or poor condition, because prevention is cheaper than reclamation. If priorities are to be established on an economic basis, it is better to concentrate first on protecting and improving the good land rather than the poor land (United Nations Conference on Desertification, 1977). For example, most of the irrigated land is not now salinized or waterlogged; much of it can be kept that way with only the application of good field water management practices. Similarly, simple range management techniques can protect the tsetse-infested savannas of Africa when they become usable for livestock production. Correcting abuses, on the other hand, usually is difficult and costly.

COMBATTING DESERTIFICATION

Combatting desertification calls for several kinds of action, some technological, some political, and some social. There is little likelihood that control measures will be instituted or made effective unless there is individual, group, and governmental will to bring about improvement. Beyond that, there may be economic, educational, and legal constraints that hamper implementation of reclamation or rehabilitation efforts. Cultural factors seldom prohibit adoption of improved practices if adequate incentives are offered and local sensitivities are given due regard. Basic principles of good land management are well known but their application to a particular situation generally requires adaptive studies and a competent advisory service to assist users.

Moreover, combatting desertification cannot be done as an isolated effort directed solely toward that end. Anti-desertification programs must be a part of a broader development program that is designed to bring about improvement of rural conditions of health, education, communication, and welfare. A single-purpose program of land improvement has little chance of success if it attacks only the symptoms of the problem (Bunting, 1979).

Land degradation is the result of past and present management practices. Since conditions are continuing to deteriorate nearly everywhere, it is obvious that coping mechanisms currently employed are inadeuqate. Change is necessary, but not change for the sake of change.

Restraints

Several factors restrain attempts to control desertification. Given man's ingenuity and a strong determination to meet the challenge, however, the possibility of circumventing the restraints appears to be good.

Any listing of factors limiting efforts to combat desertification would have to begin with the land pressures generated by increasing numbers of people and livestock. With the developing world's human population growing at an annual rate of 2.5 per cent or more, food, fuel, mineral, and recreational demands increase correspondingly. Greater affluence - largely among urban people - in addition to greater expectations of sharing in the economic wealth of a nation can indirectly place a heavy burden on finite natural resources. The global population is certain to continue growing for several decades, in the absence of a worldwide catastrophy. People probably will be more affluent in the future, on the average, and expectations or demands for a better life can be expected to go on rising.

Another of the restraints is the absence of unused productive land, except in the tsetse fly regions. By and large, the best land is already being cultivated. There are a few opportunities to expand cultivation on to grazing lands - as in the Argentine pampas - and many opportunities to irrigate potentially productive dry lands, but at a high cost.

A powerful factor working against the effective implementation of desertification control measures is the general inability of traditional agriculture to feed an expanding rural and urban population without upsetting the ecological balance. Traditional agriculture is a low-risk low-yield food production system capable of functioning indefinitely at the subsistence level if the dependent population is small and their needs modest. When the population increases and freedom of movement to other land is restricted, the system usually cannot cope with the new and unusual conditions that arise. The impact is pronounced in the arid and semi-arid regions where traditional agriculture provides only a precarious existence for the average pastoralist or cultivator.

The most readily available apparent solution to the problem of inadequate food production is to extend cultivation and grazing into increasingly less productive regions, increase livestock numbers, reduce size of land holdings, or shorten the fallow period between crops. Those measures may increase production for a few years but the long term result is accelerated land degradation and reduced productivity. The checks and balances that formerly provided some protection to land that had been abused (famine and disease, abandonment of old land and movement to new land) are no longer operative at the scale they once were. The inescapable conclusion is that production practices must be intensified if yields are to be increased to accommodate a larger agricultural and urban population (Dresch, 1977; Depiere and Gillet, 1971; Buringh and van Heemst, 1979). The potential for doing so, while at the same time stopping and reversing desertification, is great, but it will not be easy.

Intensification of production on small farms where only hand labor is available is especially difficult in dry farmed areas. Productivity is limited first by rainfall, not by soil fertility, and such practices as intercropping, multiple cropping, close spacing of plants, and heavy applications of manures or fertilizers generally are either uneconomic or counterproductive. Nevertheless, much can be done to increase production by modifying current practices rather than introducing radical changes of the kind represented by big tractors.

The difficulty most farmers and herders have in financing needed land improvements from current income severely restricts the implementation of proven desertification control practices. Furthermore, because the improved technology frequently does not lead to increased animal or crop yields until a year or more has passed, land users may be unwilling to go into debt to finance improvements, even if credit can be obtained. In the case of grazing land restoration, the need to reduce livestock numbers - as the first step in range management - means that income will be reduced for the first few years, which places a heavy - if not intolerable - burden on poor people. It is considerations such as these that lead many people to contend that preserving and reclaiming land resources is at least in part a national as well as individual responsibility.

Finally, there is a host of additional restraints that are widely recognized as important. They include the dearth of credit, low education level, inadequate advisory services, poverty, poor transportation facilities, and bureaucratic inefficiency, among others.

Role of Technology

Adoption of improved technology appears to be essential if desertification is to be controlled, given the inability of the technology in current use to prevent continued land degradation. The technology needed is, of course, that which is appropriate for the educational and financial level of the land users. By and large, the technology that is the most simple, cheap, easy to use, and effective is the most appropriate technology. It may be high-level or low-level technology, depending upon what is needed to fit the circumstances.

Optimally, the use of advanced systems of technology in agriculture would increase crop production on the better lands to the point where the poorer lands could be returned to less intensive grazing, forestry, wildlife, and recreational uses, as has been done in the humid regions of Europe. If that easing of land pressures could be accomplished, combatting desertification would be a much less difficult task. The current need to eke out as much as possible from every piece of available land - no matter what its condition - is a severe constraint on land improvement efforts.

Reliance on advanced technology to solve the agricultural problems responsible for worldwide desertification may be essential but it has drawbacks. Perhaps the most important caution to note is that technological solutions are unlikely to be permanent solutions even if they are well-planned (Peterson, 1979). New problems arise that require new solutions. There are certain to be negative as well as positive effects, and constant vigilance is necessary to keep the negative effects from getting out of control. The spread of schistosomiasis made possible by irrigation projects is but one example of the complications that can arise when even the most worthy project is undertaken.

Although the introduction of advanced technology carries with it the risk that there may be undesirable side effects, it also carries the promise of being able to feed the world - even with a greater population than that predicted for the year 2000 - while preserving the land resource. Unless agricultural productivity is raised, there is no hope of stopping desertification in land-poor countries. Since some risk is associated with the introduction of a new technology, careful planning is needed to minimize that risk. One of the best ways to minimize risk of a large scale disaster occurring is to conduct a field-scale pilot test that will bring to light unanticipated side effects. The Iraq and Pakistan case studies on desertification emphasize the need for pilot testing (Dougrameji and Clor, 1977; Irrigation, Drainage, and Flood Control Research Council, 1977). Unfortunately, pilot testing is widely seen as desirable but is seldom carried out in advance of development activities.

The goal of obtaining sustained high productivity from agricultural lands is consistent with combatting desertification, in that high productivity cannot continue if environmental deterioration is permitted to go on.

MANAGEMENT PRACTICES

Solutions to the desertification problem consist, in the field, of management actions that involve improved technologies. Since the improved technology - or at least the basic principle - probably has been available for a long time, the question can be raised of why it has not already been adopted. Several reasons can be advanced but they all come down to three: 1) lack of knowledge about the technology and how it can be used, 2) inability to implement the desired changes, and 3) lack of incentives to do so. All three reasons have validity, but the last one usually is the main one. Availability of adequate incentives is the social and political aspect of combatting desertification that determines whether technological solutions will be found and applied.

Statements to the effect that technological solutions to the desertification problem are on hand are both correct and incorrect. They are correct in that, for example, waterlogging can be avoided by providing adequate surface and subsurface drainage or by reducing canal seepage or by improving field water application methods. The statements are incorrect if they imply that there is a standard method that can be used to achieve each of those objectives or to choose among them. Only an understanding of local conditions will permit that choice to be made, whether it is based on technological, economic, social, or political considerations or some combination of them. That is why adaptive research and field testing are so essential to the success of a development project. Field testing also helps to identify and to develop a response to the perturbations in the system that will occur when a technology is introduced, as one change leads to another. The failure of an introduced technology may well be due to failure of the technologists to adapt their techniques to the local physical conditions, rather than being due to the unidentified socioeconomic conditions that are so frequently blamed for those failures.

Grazing Land

Desertification of grazing lands affects a greater part of the arid regions than any other kind of land degradation. About 90 percent of the 3,600,000,000 hectares of rangeland having more than 100 mm of precipitation are desertified to some degree. Moderate soil erosion is the normal condition on overgrazed range lands but the most important effect is degradation of the plant cover and the soil surface. The concept of an "encroaching" desert is a plant ecologist's reaction to the deterioration in vegetation that accompanies overuse (Le Houérou, 1970). As Kassas (1970) describes it, the desert encroaches upon the steppe, the steppe on the savanna, and the savanna on the forest. The encroachment consists of steppe vegetation being replaced by a desertic type of vegetation, savanna vegetation by a steppe type, and forest vegetation by a savanna type.

Each shift in a vegetation belt (e.g., savanna replaced by a grassy steppe) impoverishes the flora and substitutes more xerophytic plants for the original less xerophytic ones. In doing so, the new plant aspect gives the impression that the climate must have changed, when, in fact, it has not. Greater runoff accompanying the deterioration in plant cover and surface soil structure may have made the soil (edaphic) environment more arid but the climate has remained the same. When these shifts in ecological zones occur over a large region, as in the Sudan, the appearance is given of an advancing desert. The fact that the change need not be permanent is demonstrated by the frequently rapid recovery of the plant cover in exoclosures or where the shrub competition has been removed and grazing reduced. In some few cases, soil deterioration associated with the formation of large gullies and sand dunes does bring about permanent change.

Improvement of overgrazed range land calls for the utilization of a grazing system that allows recovery of the vegetation periodically. Various deferred-rotation management systems have been devised (Stoddard et al, 1975). Choosing among them depends upon kinds of forage present, size of area, seasonal or yearlong use, and other factors. All require control of the grazing time and intensity, something which is difficult to do on communal land unless there is strong and decisive leadership. As is true of so many introduced practices, improving the grazing system, alone, will not be successful unless accompanied by better water management and control of livestock movement by fencing or careful herding.

The singly most difficult step to take in restoring abused ranges to a high level of productivity is the first and most essential act: reduce livestock numbers. Once that reduction has occurred and the range vegetation improved, the livestock carrying capacity may become several times greater than it was in the deteriorated condition. However, the interim period when production of milk and meat are low places on the pastoralists heavy burdens that they may not be able to carry unless given outside assistance.

In the dramatically different short-duration grazing system introduced into South Africa, reduction of livestock numbers as the first step in restoring deteriorated rangelands is not necessary, in most cases (Howell, 1978). However, that system, which is said to allow an immediate doubling of livestock numbers on overgrazed land, requires expensive fencing and handling facilities and careful management. The fact that the short-duration grazing system, when conditions are properly controlled, has been successful gives hope that other management practices can be devised that will be equally successful but less expensive to install.

While some type of deferred-rotation grazing system is essential for improving and maintaining range lands, other practices may also be useful. Water spreading, water well construction, contour furrowing, range pitting, brush control, reseeding, and fertilization are among the most effective. There is no formula for choosing among the alternatives; experience and pilot testing provide the best guidance.

Combatting desertification of grazing lands has had a modicum of success in the United States since the 1930's although much remains to be done (Table 5.1). Part of the improvement can be attributed to more favorable moisture conditions following the 1930's drought but another part certainly is the result of better range management.

TABLE 5.1. Range condition trends in federal lands of western United States*

Trend	1930-35 %	1975 %
Improving	1	19
Stable	6	65
Declining	93	16

*Adopted from Hadley (1977).

Watering Points

The distribution of watering points controls the utilization of range forage. Despite the much-maligned provision of new and improved wells in the Sahel before and during the 1969-1973 drought and the contention that they aggravated land degradation, water management is an excellent tool in grazing management. It must be done carefully, however, as is true of any other management practice. Indiscriminate drilling of wells is no panacea for improper range utilization, and the congregation of livestock around watering points can pose difficult management problems. It is much easier and more popular to dig wells than to control livestock numbers and movement.

Goats

Suggestions are regularly made to ban the goat in the arid regions, on the premise that the goat is responsible for wholesale destruction of woody plants. That recommendation ignores the fact that the goat becomes the preferred animal when the grazing environment is too poor - naturally or due to man's activities - to support cattle and sheep. Goats certainly are destructive of shrubs and trees but their presence on what was once good grazing land generally is symptomatic of the desertification problem rather than the cause (Hernandez X., 1970). Their ability to climb trees and convert spiny vegetation into meat and milk makes them important for the sustenance of nomads and poor farmers intent upon survival in hostile environments (Cloudsley-Thompson, 1970).

A re-evaluation of the role of goats in a pastoral economy has shown that they can be useful in a well-managed cattle-goat grazing system (Merrill and Taylor, 1976). Goats help to control the woody vegetation that competes with the herbaceous vegetation that cattle (and sheep) prefer, which reduces or eliminates the need for expensive and environmentally undesirable brush control measures. Collective grazing of animals whose food habits complement each other offers an opportunity for increasing rangeland productivity (Provenza, 1978). The principle is the same as for game ranching, where trees, shrubs, and grass are consumed selectively by different species.

Woodcutting

There is little likelihood that the destructive cutting of trees and shrubs for firewood and charcoal will be stopped until a cheap substitute is found or tree plantations are established. At present there is no cheap substitute fuel available in the arid regions. Much effort by governments and private agencies has gone into reforestation projects designed to reduce soil erosion, increase soil fertility, and provide wood on a sustained-yield basis but those efforts have met with little success, thus far.

The costliness or unavailability of wood and kerosene in developing countries increases the use of animal dung for fuel instead of for improving soils. If the United States experience in its arid regions is any example, woodcutting and dung burning will continue to deplete those resources - and encourage desertification - until a better substitute for wood fuel is found. There is none in sight even though the potential for solar cookers and heaters in the cloudless arid regions appears to be unlimited. Tree plantations are an obvious solution but they have been largely ineffective due to seedling establishment problems and the need to guard the trees continually against premature cutting to meet immediate requirements (Thomson, 1977).

Fire

Grass fires have been blamed for degrading the vegetative cover and for wasting forage that livestock could utilize. It is true that indiscriminate burning can have an adverse effect on plant and animal life, but it is also true that carefully controlled burning is a useful management tool. Fires can serve not only to remove tough and unpalatable old stems of plants and stimulate new and succulent growth but also to control shrub and tree growth (Ramsay and Innes, 1963).

Burning is the only effective tool available in developing countries to keep land relatively free of encroaching woody vegetation and to bring about other short-term benefits. Prevention of burning encourages the invasion of grasslands by Acacia, Prosopis, and other woody plants (West, 1978).

Sand Dune Stabilization

A variety of measures have been used successfully to stabilize sand dunes. Petroleum (asphalt, crude oil) mulches are effective in immediately stabilizing shifting dunes (Forestry and Range Organization, 1977) but they are expensive to use because of the equipment required to heat and spread the mulch material. Latex sprays are sometimes used to accomplish the same purpose as petroleum mulches. More commonly, dead vegetation and fences are used to stabilize dunes until live vegetation can be established. A checkerboard or linear pattern of windbreaks is customarily employed, the former when winds blow from several directions and the latter when winds are dominantly unidirectional.

Stabilizing sand dunes, wherever possible, with growing trees, shrubs, and grasses is the best means of achieving long-lasting control of sand movement. The presence of a moist layer at depths below 50 cm in some sand dunes can be used to establish plants without irrigation. The moist layer occurs as the result of the self-mulching character of sandy soils (Abdel Rahman and

El Hadidy, 1958). Rapid drying of the surface sand after a rain produces a mulch effect that greatly reduces further evaporative losses and allows water to be retained in the lower layers. Planting *Tamarix* cuttings in the moist layer of shifting sand dunes has been used effectively in eastern Saudi Arabia to stabilize the dunes. Artificial mulches and wind barriers are useful for protecting young plants from sand-blasting and exposure of roots while they are becoming established, but the effects are generally short-lived. Thin petroleum mulches may provide protection for two or three years - if the mulch is not broken by animal or human traffic - before they must be renewed; thick mulches can last longer.

Despite the considerable amount of information on sand dune stabilization, field tests are advisable before large-scale projects are undertaken. Among the many variables that can affect stabilization efforts are the availability and salinity of irrigation water used to establish plantings, adaptability and growth characteristics of plant species, wind direction, steepness of slopes, depth to groundwater, mobility of the dune, suitability of machinery for transplanting and mulch application, productivity of nurseries, and a host of other factors. An excellent discussion of virtually all aspects of sand dune stabilization can be found in Hagedorn et al (1977).

Mining

Restoration of surface mined land to something approximating the original condition is at least as difficult as restoring overgrazed range land that has been rendered essentially useless by water and wind erosion. As in range lands, the problem of revegetating mine spoils becomes greater as the climate changes from semi-arid to extremely arid. What can be done easily in the wetter areas by simply smoothing the land, spreading stored topsoil on top of the spoils, and re-seeding with native or introduced species may be very difficult or impossible to accomplish in the deserts unless irrigation water is supplied permanently (Thames, 1977; Wright, 1978).

In addition to being infertile and easily erodible, coal mine spoils may be quite saline, whereas hard rock (copper, etc.) mine spoils frequently contain various heavy metals that are toxic to plants. Revegetating spoils of those kinds is especially difficult, even in the semi-arid regions, and usually requires covering the spoils with a thick layer of good soil.

The bare spots surrounding the sites of oil and gas wells and openings to underground mines cannot be revegetated easily as long as the well or mine is in operation. Wheeled traffic is bound to destroy vegetation. About all that can be done is to keep the affected area as small as possible.

Recreation

Tourists in the arid regions have the capacity to enjoy the natural beauty and resources of the region or to destroy them. Off-road vehicles are almost always destructive once they leave established roads. On the

PREVENTION AND REVERSAL OF DESERTIFICATION   153

other hand, visitors who obtain their pleasures from looking for attractive rocks, exploring strange places, taking pictures of the unique plants and animals, prospecting for minerals, riding horses or camels, sailing on clay flats, or just watching spectacular sunsets can enjoy themselves while preserving the natural resources, in most cases. The latter type of visitor can be an asset; the former is a threat.

The objective of recreational use of arid lands should be to preserve the natural resources while permitting their enjoyment with as few restrictions as possible. The challenge is to safeguard archeological, scenic, biological, or historic sites, to protect critical watersheds, and to assure public safety in hazardous areas without having endless regulations governing land use (Off-Road Vehicle Advisory Council, 1970). An educational program can help to make visitors aware of the need for individual and collective action to protect the resource from abuse. Beyond that, some rules and regulations on land use probably will be necessary. In cases where land abuse by off-road vehicles cannot be effectively regulated over large areas, it may be desirable to set aside small "sacrifice" areas where land degradation could be tolerated. Recreational use of off-road vehicles could then be confined to the sacrifice areas.

Problems associated with crowds numbering in the thousands gathering to capture the beauty of a sunset, as at Ayers Rock in central Australia, are also difficult to solve. Perhaps the answer is the same as for off-road vehicles: accept the desertification of a small area in order to protect the remainder. Unless a sacrifice area is delimited, it becomes necessary to control the number of persons and vehicles allowed to use the land. As is so often true, small numbers of people can be accommodated; large numbers pose problems even when the impact of one individual is of little consequence.

Dryland Farming

Land degradation in the rainfed cultivated areas lying within the dry zones of the world is the result of water erosion, wind erosion, loss of soil fertility, soil crusting, soil compaction, and salinization. Techniques are known by which all of these problems can be controlled. The most difficult ones are soil crusting and wind erosion.

Dryland farming is practiced on about 225,000,000 hectares of land, nearly 80 percent of which has suffered some desertification. Water erosion, with the associated loss in soil fertility and deterioration in soil physical condition, is the major problem on sloping land and wherever rains are of a torrential nature. Wind erosion occurs throughout the dry farmed area but is most acute on the drier fringes, especially during droughts. The most obvious symptom of desertification, worldwide, is the reduction in soil fertility that shows up in the appearance of the crop and in reduced yields. Soil crusting is the result of a loss of soil organic matter due to cultivation and of a lack of vegetative matter to protect the soil from the impact of raindrops. Subsoil compaction comes about when the weight and pressure of tractors and tillage implements increase soil density and reduce porosity. Man-made salinization is a problem which apparently is serious only in relatively small areas in North America and Australia, although conditions in wheat regions of subhumid China and the U.S.S.R. would indicate that the same problem may be present there.

Water erosion

Water erosion control consists of combining a number of land management practices with runoff control structures. Terraces of the bench type have long been used on steep slopes in China and Latin America. In China, their purpose was to reduce erosion; in at least part of the Andean zone of South America, terraces were constructed to improve irrigation water control and to catch eroding sediment from higher slopes and build up a productive soil on the terraces (Donkin, 1979). Several variations of bench terraces have been employed over the past thousands of years by early societies in the Americas, Asia, Europe, and Africa as a means of capturing sediment and water and of facilitating irrigation. In more recent years, research on erosion control and water conservation has led to the development of broad-based, narrow-based, conservation-bench, and other kinds of terraces that are usable on flatter slopes and do not require levelling the land between terrace embankments (Beasley, 1972). Terracing on the contour, once considered indispensable for erosion control, has been replaced in the United States by parallel terracing, which is better adapted to mechanized agriculture and is effective for erosion control. Guides for fitting terracing to practically any set of circumstances have been developed.

Terracing should never be the only erosion control practice. To be most effective, it should be combined with other practices such as minimum tillage and strip cropping. A vegetative cover is essential to reduce splash erosion and soil crusting.

Wind erosion

Wind erosion control is considerably more difficult than water erosion control. One reason is that maintaining a plant cover in dryland areas is dependent upon having enough crop growth to produce adequate plant residue to protect the soil. If rainfall is deficient or badly distributed in time, crop yields - and residue - are reduced. When a severe drought strikes, crop failures and dead windbreaks leave soils unprotected just when winds are likely to be strongest.

Controlling wind erosion is almost entirely a matter of utilizing vegetative matter to do so. Plant residues in the field and windbreaks (shelterbelts) in or between fields are the principal techniques in use. Fences of various kinds, solid walls, asphalt surfacing, and clay surfacing are employed mainly as means to stop sand dune encroachment on oases, railroads, and highways. These latter techniques are expensive to construct and expensive to maintain.

For whatever reason, wind erosion control seems to be a modern development although the value of windbreaks has been long known. In the United States, it was not

PREVENTION AND REVERSAL OF DESERTIFICATION

until after the Great Plains were cultivated that attention was paid to reducing wind damage on a large scale. Several tillage implements were devised to leave plant residue on the surface (stubble mulching) and to turn up clods that would roughen the surface. Minimum tillage is designed to provide the maximum amount of plant cover on erosion-susceptible land, and the perfection of that system of land management offers the greatest promise of immediate reduction in soil erosion.

The Soviet Union and the United States, both of which have extensive areas of grasslands that have been converted to dryland farming, undertook massive shelterbelt projects in the 1930's. Shelterbelts seem to have been accepted better in the Soviet Union than in the United States, where many of the early windbreaks have been removed. Cultivators objected to eight and ten row shelterbelts because they took relatively large areas of land out of cultivation and because tree roots sapped the adjoining cultivated land of water and nutrients. In recent years, another objection has been raised: shelterbelts interfere with the movement of center pivot sprinkler systems. Shelterbelts have been better received in the cooler northern Great Plains, where snow entrapment as well as erosion control are benefits, than in the hotter southern Great Plains. In the south, water absorption by tree roots is more objectionable and establishment of the required shrubs and trees is more difficult.

Replacement of eight and ten row shelterbelts by two row plantings has increased their acceptability. At the same time, techniques for improving plant establishment have been developed and the cost of installing a shelterbelt has been reduced. Research on field windbreaks, in which one or two rows of a fast growing tall grass or crop are placed between several rows of a short erosion-prone crop, offers promise for the future.

Plant Nutrients

Soil fertility losses, principally of nitrogen, can be replaced by turning under soil improvement crops, by intercropping with legumes, or by applying manures and commercial fertilizers. These measures are most economic in the wetter part of the dry farmed zone and least likely to be profitable in the drier part.

Soil Compaction and Crusting

Soil compaction is a problem that is confined to fields where mechanized agriculture is practiced. The control of the problem consists of changing tillage practices to vary the depth of tillage.

Soil crusting is difficult to reduce because the principal cause of crusting is the reduction in organic matter that almost inevitably accompanies dryland farming. Protecting the soil surface from the direct impact of raindrops is an effective control method but the amount of plant residue available to do so may be too little in dry years.

Irrigated Land

Salinization, waterlogging, and soil compaction are

the principal features of land degradation in irrigated areas. They are responsible for reducing crop yields on about 20 percent of the 130,000,000 hectares of irrigated land in the arid regions. Techniques for controlling those problems are well known and fairly simple to use if soils are permeable. If soils are rather impermeable, the task becomes harder and requires a considerable degree of technical knowledge. Fortunately, reducing damage from these soil problems also increases water use efficiency and crop yields at the same time, which helps to make the soil improvement immediately profitable. Salinization and waterlogging are much more serious, worldwide, than soil compaction.

High water tables lead to soil salinization nearly everywhere in the arid regions. Since drainage systems rarely are constructed when the irrigation system is, waterlogging usually shows up several years after irrigation has begun. If the waterlogging and subsequent salinization generate enough concern, a drainage system will be installed. After that, the long-time success of the irrigation project will depend upon how well the land is levelled, the kind of water management that is practiced, and the maintenance of the irrigation and drainage system. About the only places where drainage problems can be expected to not occur sooner or later are those where natural internal drainage prevents the accumulation of groundwater or where pump irrigation is practiced.

Waterlogging is controlled by improving water management and by providing underground drainage. Salinization control calls for leaching the soil and removing the drainage water. The special case of sodic soil reclamation requires a treatment that replaces adsorbed sodium with calcium, followed by leaching and drainage water removal.

Soil compaction is significant only where mechanized agriculture is practiced. Varying the depth of tillage breaks up compacted subsoil layers. Compaction caused by heavy machinery can be reduced by modifying the machinery to spread the weight, by cutting down the number of trips across the land, or by driving machinery over only a part of the field.

DESERTIFICATION CONTROL

The fight to stop, then reverse, desertification is very difficult when a nation is mostly arid and poor in resources. Countries having significant areas of humid climate within their borders, strong industrial bases, or high-demand resources such as oil are the fortunate ones. They are often capable of generating funds to pay for ameliorating measures. Rich nations can use high-cost technology or expensive subsidy programs to combat desertification; poor nations cannot. Poor nations, therefore, must resort to other measures; improvements will be slow and require long-term commitments.

In the world of today, land pressures in the arid regions have had their greatest impact on range and rainfed cropping lands. Pastoralists have increased their herds following a number of mutually-supportive measures taken by governments which have had particularly significant effects in increasing livestock numbers. These measures include improved veterinary services, better water supplies, decreased animal predation, and increased security against raiders. Individually, each of these measures is desirable. Unfortunately the successful combination of these factors produces increased livestock numbers on each unit of land but without any improvement in management techniques.

The complexity of introducing range management techniques is well illustrated by the quandry of pastoralists in developing countries. Obviously, the number one range management need is to reduce livestock numbers in overgrazed areas. This need is contrary to the pastoralist's own experience which has proven to him that survival in times of stress depends upon having as large a herd of livestock as possible. The more animals he has at the beginning of a drought, the better are his chances of surviving on his own resources. As long as the only way to accumulate capital is by investing in cattle, camels, goats, and sheep, no prudent pastoralist will do otherwise. The harmful effect such practices have on vegetation is obvious. But, it is equally obvious that reduction in grazing pressure will not occur unless other powerful incentives are introduced.

Population increases have pushed sedentary farming into drier and drier areas where crop production risks are great and fallow periods during which land could recover from cropping have been shortened. Governments have in other instances encouraged growth of cash crops on better land and left poorer land for local food production. The net result is cultivating more marginal land and allowing less time for fertility recovery.

Fire, goats, and woodcutting have been widely blamed for causing environmental deterioration. And indeed they do. What is overlooked, however, is that people resort to uncontrolled burning of grass, expanding goat herds, and overcutting trees only when they are forced to. By that time desertification already is well advanced and last-ditch measures are needed to survive under increasingly adverse conditions. Edicts to ban fire, goats, and woodcutting may help halt desertification. But edicts only increase misery unless alternate food and fuel sources are provided. One of the world's greatest needs in the battle against desertification is finding cheap and acceptable substitutes for wood for heating and cooking. Reforestation of abused land will never succeed as long as people must use every last stick of wood for fuel.

A program to combat desertification should have at least four components. The first is a national commitment to initiate the task and to carry it through the completion. Along with that commitment should go the development of a national policy that will mobilize the societal will and integrate the several needed actions into a coherent whole. Only when a firm decision has been made to see the job done can significant progress be made. Emphasis on combatting desertification alone is not enough to assure continuing success. It must be seen as a part of broader national development efforts.

The second component consists of directing rehabilitation efforts at the most favorable areas. Prospects for success in combatting desertification are greatest in areas presently most intensively farmed (dryland or irrigated) where cost-benefit ratios are lowest. Starting the battle in badly deteriorated low-acre-value grazing lands would be a costly, long-term effort with modest returns. Morale of people involved would suffer because of meager impacts on the national economy. Beginning with better-endowed areas, on the other hand, would increase chances for success and would have greater immediate impacts on the economy, if successful. However favorable the setting, there is little chance of success unless the local people are involved in the planning and execution from the beginning.

Developing small-scale (cottage) agriculture-based industries is the third essential component of a successful program to combat desertification in poor nations. Expansion of the virtually nonexistent industrial base is needed for the development of nearly every nation. Local artisans are the logical persons to manufacture the equipment, starting small and learning by doing. As technological improvements become increasingly complex in future years, a labor force qualified to build and maintain equipment can be developed gradually.

The fourth component is related to general improvement of the nation's economic and education base. It calls for expanding the transportation network, improving education and applied research systems, establishing effective credit institutions, providing price incentives, encouraging better marketing procedures, and providing other infrastructure aids to crop and livestock production in an environmentally acceptable manner.

Given their inability to mount a massive campaign to combat desertification on a broad scale, poor nations may be well-advised to adopt a strategy of selective development. That tactic seeks to use the experience gained and the higher level of economic activity attained during development of better-endowed regions to gradually expand the development program to less-favored regions. The "gradualist" approach makes good sense in a nation lacking trained people, good transportation and communication networks, social services, effective credit and marketing institutions and research capability. It carries the hazard of alienating politically strong groups in areas not chosen for immediate development. And there is always the possibility that donor agencies or nations will undercut this approach by insisting on working in an area of their own choice. Selective development also does not support the recent international emphasis on improving the lot of the "poorest of the poor."

Successful implementation of programs to combat desertification has been done - on a small scale - all over the world. There is no question whether it is possible to stop and reverse desertification. The ameliorative techniques and the basic understanding of the problem are known although there is much opportunity to improve the practices in use. One of the most optimistic aspects of the desertification problem is that very little land, relatively, has been irreversibly desertified. There is still ample opportunity to increase land productivity.

Strategies for combatting desertification involve mobilizing the societal will and the widsom of the people, centering initial attention on places where prospects for success are best, developing an agroindustrial base, and strengthening indigenous institutions.

CONCLUSION

Combatting desertification is seldom easy, primarily because long-established habits of land management are difficult to change. The task becomes more difficult as the productivity of the land decreases. It is most easily done in irrigated areas where high acre-value crops are grown and is most difficult to do in the drier rangelands where plants are sparse and growth is slow.

Reclamation of salinized and waterlogged irrigated land calls for leaching of salts and the lowering of the water table. Underground drainage is required, along with land levelling to provide uniform leaching. The cost of reclamation may be high but the economic benefits are also high. Reclamation can be accomplished in a matter of months after the proper conditions have been established.

Reversing desertification in dryland farming areas generally is a slow process. The rapidity with which it can be done depends upon whether the land degradation consists of a loss of soil fertility (due to cropping and sheet erosion) or of the formation of gullies and sand dunes. The former type of degradation can be corrected in a short time by the application of fertilizers; the latter requires many months or years to control by vegetative means. Erosion control is easier to achieve in the wetter areas than in the drier locations because vegetative protection can be better provided when moisture conditions are favorable. Construction of terraces and windbreaks is costly. Conserving enough crop stubble to protect the soil against erosion is not easy to do when livestock graze harvested fields.

Improving deteriorated rangelands is a very slow

process in the drier regions and one that normally requires several years in the wetter regions. If the annual precipitation is below average, revegetation of rangelands is delayed; if above average, range improvement is hastened. Natural re-establishment of a grassland or open savanna on an overgrazed and shrub-infested range may never occur through the control of grazing, only. Some kind of shrub-control measure, with or without reseeding, will usually be necessary. Grazing management is most effective in improving rangelands when range deterioration is moderate and a good supply of seeds or plants of desirable species is present.

Restoring desertified mining and recreational areas is similar to restoring rangelands: difficult in the more arid areas and easier in the less arid areas. As with rangelands, the cost of land reclamation is high if a quick change is desired.

Preventing desertification is easier and cheaper than restoring desertified land to its most productive level. Unfortunately, the temptation to fully exploit good land commonly leads to overexploitation, at least partly because early warning of land degradation is difficult to see when growing conditions are favorable. Sensitive indicators of land condition trends are needed for irrigated, dry farmed, and grazing lands to provide that early warning of unwanted changes.

Combatting desertification requires an individual or community effort and the expenditure of money and labor. The job must be done locally although support may be obtained from outside sources. Since success cannot possibly be achieved without local action, it is essential that the affected population participate in the planning as well as the execution of the project. Much lip service has been given to that principle but local participation in planning activities seldom occurs in practice, not because the planning managers do not agree with the principle but because it is easier and faster to proceed without lengthy local consultations.

In addition to local participation in planning, there usually is a need to remove whatever obstacles there may be to the application of existing knowledge (inadequate numbers or training of technical advisory personnel, limited stocks of fertilizer and seeds, etc.). And nearly always there is a need for adaptation of existing knowledge to local conditions. Emphasis should be placed on solving long-term problems rather than on instituting short-term measures. Mistakes will be made but that should not deter efforts to learn by doing.

REFERENCES

Abdel Rahman, A. A. and El Hadidy, E. M. 1958. Observations on the water output of the desert vegetation along Suez Road. Egyptian Journal of Botany. 1: 19-38.

Beasley, R. P. 1972. Erosion and Sediment Pollution Control. Iowa State University Press, Ames, Iowa. 320 p.

Bunting, A. H. 1979. Science and technology for human needs, rural development, and the relief of poverty. IADS Occasional Paper, International Agricultural Development Service, New York. 12 p.

Buringh, P., and van Heemst, H. D. J. 1979. An estimation of world food production based on labour-oriented agriculture. Centre for World Food Market Research, Agricultural University, Wageningen,

The Netherlands. 26 p.

Cloudsley-Thompson, John L. 1970. Animal utilization. In H. E. Dregne (editor), Arid Lands in Transition, American Association for the Advancement of Science, Publication No. 90, pp. 57-72.

Depierre, D., and Gillet, H. 1971. Desertification de la zone Sahelienne au Tchad. Bois et Forêts des Tropiques, 139: 3-25.

Donkin, R. A. 1979. Agricultural Terracing in the Aboriginal New World. University of Arizona Press, Tucson, Arizona. 196 p.

Dougrameji, J. S., and Clor, M. A. 1977. Case Study on Desertification. Greater Mussayeb Project, Iraq. United Nations Conference on Desertification A/CONF. 74/10, 102 p.

Dresch, J. 1977. The evaluation and exploitation of the West African Sahel. Phil. Trans. Royal Society of London, B. 278: 537-542.

Forestry and Range Organization. 1977. Petroleum Mulch in Sand Dune Fixation. Forestry and Range Organization Ministry of Agriculture and Rural Development, Tehran, Iran, 11 p.

Hadley, R. F. 1977. Evaluation of land-use and land-treatment practices in semi-arid western United States. Phil. Trans. Royal Society of London, B. 278: 543-554.

Hagedorn, H., Griessner, K., Weise, O., Busche, D., and Grunert, G. 1977. Dune Stabilization. Deutsche Gesselschaft fur Technische Zusammenarbeit, Eschborn, West Germany, 194 p. + bibliography.

Hernandez Xolocotzi, Efraim. 1970. Mexican experience. In H. E. Dregne (editor), Arid Lands in Transition, American Association for the Advancement of Science, Publication No. 90, pp. 317-343.

Howell, L. N. 1978. Development of the multi-camp grazing systems in the southern Orange Free State, Republic of South Africa. Journal of Range

Management 31: 459-465.

Irrigation, Drainage, and Flood Control Research Council. 1977. Case Study on Desertification. Mona Reclamation Experimental Project, Pakistan. United Nations Conference on Desertification A/CONF. 74/13. 120 p.

Kassas, M. 1970. Desertification versus potential for recovery in circum-Saharan territories. In H. E. Dregne (editor), Arid Lands in Transition, American Association for the Advancement of Science, Publication No. 90, pp. 123-142.

Le Houérou, H. N. 1970. North Africa: Past, present, future. In H. E. Dregne (editor), Arid Lands in Transition, American Association for the Advancement of Science, Publication No. 90, pp. 227-278.

Merrill, L. B., and Taylor, C. A. 1976. Take note of the versatile goat. Rangeman's Journal 3: 74-76.

Off-Road Vehicle Advisory Council. 1970. Operation ORVAC. Bureau of Land Management, U.S. Department of the Interior, Sacramento, California. 40 p.

Peterson, Russell W. 1979. Impacts of technology. American Scientist 67: 28-31.

Provenza, Fred D. 1978. Getting the most out of blackbrush. Utah Science, December 1978, pp. 144-146.

Ramsay, J. M., and Innes, R. Rose. 1963. Some quantitative observations on the effects of fire on the Guinea savannah vegetation of northern Ghana over a period of eleven years. African Soils VIII: 41-83.

Stoddart, Laurence A., Smith, Arthur D., and Box, Thadis W. 1975. Range Management, Third Edition. McGraw-Hill Book Company, New York. 532 p.

Thames, John L. (editor). 1977. Reclamation and Use of Disturbed Land in the Southwest. University of Arizona Press, Tucson, Arizona. 362 p.

Thomson, James T. 1977. Ecological deterioration: Local-level rule-making and enforcement problems in Niger. In Michael H. Glantz (editor), Desertification, Westview Press, Boulder, Colorado, pp. 57-79.

United Nations Conference on Desertification. 1977. Economic and Financial Aspects of the Plan of Action to Combat Desertification. United Nations Conference on Desertification A/CONF., 74/3 Add. 2. 21 p.

West, Neil E. 1978. The changing woodlands of the Great Basin. Edge 2: 13-16.

Wright, Robert A. (editor). 1978. The Reclamation of Disturbed Arid Lands. University of New Mexico Press, Albuquerque, New Mexico. 196 p.

6. OCCURRENCE OF DESERTIFICATION

Desertification affects nearly all of the arid regions, to varying degrees, except for the extremely arid climatic deserts such as the Sahara, Atacama, and Takla Makan. An indication of the extent and intensity of the desertification that has occurred in the past is essential if there is to be an appreciation of the importance of the problem. It should be recognized that delineating desertification on the small scale continental maps in this chapter cannot provide anything more than an indication of the kind and the magnitude of the problem; it cannot show accurately the size of small affected areas.

The classification system used in the preparation of the continental desertification maps is based on four classes of desertification: slight, moderate, severe, and very severe. The criteria for each class are as follows:

Slight: Little or no degradation of the soil and plant cover has occurred.

Moderate: 1) 26 to 50 percent of plant community consists of climax species or 2) 25 to 75 percent of original topsoil lost or 3) soil salinity has reduced crop yields 10 to 50 percent.

Severe: 1) 10 to 25 percent of plant community consists of climax species or 2) erosion has removed all or practically all of the topsoil or 3) salinity controllable by drainage and leaching has reduced crop yield by more than 50 percent.

Very severe: 1) Less than 10 percent of plant community consists of climax species or 2) land has many sand dunes or deep gullies or 3) salt crusts have developed on very slowly permeable irrigated soils.

The "very severe" category represents the extreme condition that many people associate with desertification. It is land so badly degraded that its utility by man or animals is virtually zero and the degradation is economically irreversible, for most purposes. While there are many small areas of land that fit into this category, there are few areas large enough to be shown on the continental maps. Practically all of the world's desertification can, at this point, be reversed.

Delineations on the maps, because of the small scale employed, usually are combinations of different desertification classes. For example, an area shown as moderately desertified may have inclusions of slightly desertified and severely desertified land. The guidelines established to determine the map classification for any delineated area are given in Table 6.1.

TABLE 6.1. Desertification classification criteria

Map classification	Percent of area in various desertification categories
Slight desertification	>50% of area in slight category <20% in severe category <10% in very severe category
Moderate desertification	<50% in slight category <30% in severe and very severe category
Severe desertification	>30% of area in severe category 0-30% of area in very severe category
Very severe desertification	>30% of area in very severe category

Table 6.2 gives data on the land area of the arid regions of the world in the four desertification classes. Most of the land used for agriculture in the arid regions is at least moderately desertified. The 52.1 percent figure for land in the category of slight desertification consists mainly of the naturally barren climatic deserts (hyper-arid regions) where man's impact has been minimal. Worldwide, about 21 percent of the irrigated land, 77 percent of the rainfed cropland, and 82 percent of the rangeland are at least moderately desertified (Table 6.3).

TABLE 6.2. Desertification of arid lands of the world

Desertification class	Area affected	
	Square kilometers	Percent of arid lands
Slight	24,520,000	52.1
Moderate	13,770,000	29.2
Severe	8,700,000	18.5
Very severe	73,000	0.2
Total	47,063,000	100.0

AFRICA

The severe Sahelian drought that extended from 1969 through 1973 focussed the world's attention on the human aspects of land degradation and led to the convening of the 1977 United Nations Conference on Desertification. Drought, however, is not the cause of desertification; man is. The drought served only to place additional stress on the biological resources of the Sahel. If resource management had been good, little, if any, permanent damage is done by droughts. However, if resource management has been unwise, a drought accentuates the adverse impact of that management and accelerates land degradation (Weaver and Albertson, 1940). The latter is what has occurred widely in the Sahel and elsewhere.

All of the usual forms of desertification are present in the arid regions of the African continent and are manifested as serious local or regional problems (Figure 6.1). Overgrazing has reduced range productivity

TABLE 6.3. Arid lands affected by desertification

Continent	Irrigated Land			Rainfed Cropland			Rangeland		
	Total	Area affected by desertification	%	Total	Area affected by desertification	%	Total	Area affected by desertification	%
	------000ha------			------000ha------			------000ha------		
Africa	7,756	1,366	18	48,048	39,633	82	1,182,212	1,026,758	87
Asia	89,587	20,572	23	112,590	91,235	81	1,273,759	1,088,965	85
Australia	1,600	160	10	2,000	1,500	75	550,000	307,000	56
Europe (Spain)	2,400	890	37	5,000	4,200	84	16,000	15,500	97
North America	19,550	2,835	14	42,500	24,700	58	345,000	291,000	84
South America	5,389	1,229	23	14,290	11,859	83	384,100	319,380	83
	126,282	27,052	21	224,428	173,127	77	3,751,071	3,048,603	81

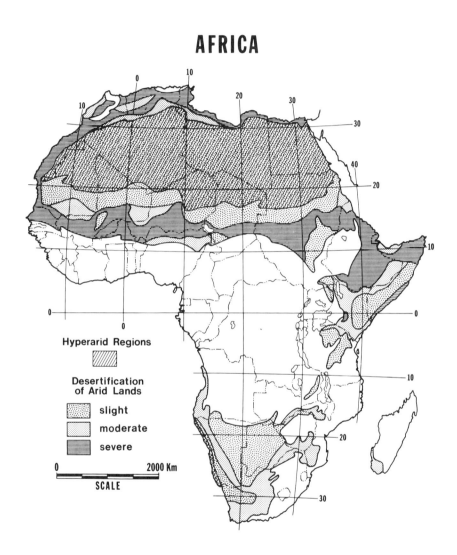

Figure 6.1. Status of desertification of arid lands of Africa

virtually everywhere outside the tsetse fly regions, in north, west, east, and south Africa. Wind and water erosion have devastated landscapes in the cultivated regions and in much of the rangelands (Rapp, 1974). Shortened fallow periods in the shifting cultivation system south of the Sahara have led to severe depletion of plant nutrients. Salinization and waterlogging of irrigated land is worst in the Nile Valley and North Africa but also occurs elsewhere. Mining has left surface scars wherever it is practiced. The environmental degradation continues and shows no sign of slowing down.

Desertification Characteristics

A number of factors have increased land degradation and the vulnerability of the African arid regions to desertification. Most of them have had similar effects in Asia and Latin America. They can be grouped in three categories: 1) increased human and animal population, 2) improved health services, and 3) injudicious use of technology.

Due to the increased sedentary population, pressures on cultivated land led to a shortening of the fallow period in the shifting cultivation cycle and the extension of cropping into the more precarious drier regions. Soil fertility declined in the less arid regions, wind and water erosion increased, and crop harvests became less reliable and more variable as the desert edge was approached. Concurrently, nomadic pastoralists were deprived of some of their best grazing lands as the cultivators moved in (Delwaulle, 1977; Floret and Hadjej,

1977). At the same time that the rangeland area was contracting, populations of pastoralists and their livestock were increasing and the provision of improved veterinary services and the lack of a viable marketing system helped assure that animal numbers would grow rapidly (Widstrand, 1975). The result was inevitable: overgrazing and accelerated desertification.

In the Sahel, shifting cereal cultivation from field to field, with fallow periods of 5 to 10 years, or moving entire villages to new areas every 20 to 30 years provided the rest period needed to restore soil fertility, in years past. As population density increased, fallow periods were shortened or even eliminated, especially close to the villages (Depierre and Gillet, 1971). Cultivation was also extended farther north, into the drier zones. The government of Niger once promulgated a law forbidding dryland cropping north of the 14th parallel, but the law could not be enforced. In Tunisia, consideration has been given to enacting similar legislation that would prohibit cereal cultivation in the more climatically hazardous regions of the south. Floret and Hadjej (1977) contend that neither legislation nor public education would stop the spread of cereal farming; alternatives must be provided that will improve the economic status of the agriculturalists. Further, they say that any attempt to combat desertification is doomed to failure unless it is integrated into an overall program of socioeconomic development.

Surface runoff was reduced significantly in Uganda when shrubs were removed and regeneration with grass was accomplished (Pereira, 1977). Depth of water penetration was tripled and stock-carrying capacity was doubled on the improved watershed. The degree to which surface crusting affected runoff was not evaluated.

Overgrazing inadvertently was made worse, particularly in the Sahel, by the drilling of additional wells that provided drinking water for livestock throughout the year. Without the rest period that intermittent water supplies previously assured, forage conditions deteriorated around the wells where water was no longer a limiting factor in livestock survival. Local authorities did not or could not impose a control system that would allow forage plants to recover from heavy grazing.

A crude estimate of the amount of crop yield reduction that could be attributed to salinization and waterlogging in the Nile Valley indicates it to be at least 17 percent on 80 to 90 percent of the old irrigated land. In the newly irrigated lands east and west of the delta the figure is likely to be a minimum of 25 percent yield reduction. Both problems are serious ones that can become even more serious in the future if remedial measures are not taken. The situation in Algeria and Tunisia is similar to that in Egypt with respect to yield reductions.

Solutions to desertification problems in Africa are known and - in general - can be implemented readily if resources are available to do so. An exploding population in the developing countries, however, means that land pressures will continue to build. Reducing livestock numbers in the grazing lands until forage productivity can be improved is very difficult, at best, and introducing range use controls on communal land is not easy. There is little likelihood that marginal dryfarm lands will be returned to pastoral use unless production on the better lands is increased enough to offset the loss of cultivated land. While the latter can be done, progress in that direction is very slow.

The importance of controlling salinization and waterlogging in Egypt and North Africa has been recognized for many years. Numerous projects have been undertaken to improve soil and water management, which is the key to the problem. Prospects for improving the old irrigated lands in the Nile Valley are promising but development of the new lands presents a continuing challenge.

Destruction of woody vegetation has been hastened by the ever increasing need for firewood to meet the demands of the larger population. The destruction is especially noticeable around the rapidly growing urban centers, where the circle of deforested lands gets larger every year (Delwaulle, 1973).

One of the consequences of woodcutting in arid lands is said to be falling water tables. This is true in the "arid" climatic region on the Unesco map of arid regions but not in semi-arid regions. In semi-arid regions, for the most part, cutting down trees raises the water table instead of lowering it. This is certainly true in the savannas (Grove, 1974; Charreau, 1974). The phenomenon is the same as that responsible for saline seeps in Australia and North America: destruction of trees reduced transpiration and increases the amount of water that percolates downward.

While desertification was a long-standing problem even in the absence of droughts, the gradually increasing vulnerability of the land made the impact of the inevitable droughts worse than ever (Dahl and Hjort, 1979; Depierre and Gillet, 1977). The factors responsible for that vulnerability are still operating, desertification continues, and future droughts will have ever greater damaging effects.

Salinization and waterlogging have plagued the Nile Valley for decades (Aboukhaled et al, 1975), principally on the perennially irrigated land. The problem is caused by a combination of excessive use of irrigation water, seepage from canals, absence of adequate drainage, and the presence in places of highly mineralized ground water (Elgabaly, 1969). The quality of the Nile River water is excellent.

In the Sudan, the clay plains of the Gezira and neighboring areas are sometimes sodic - according to soil tests - but crops are not adversely affected. That curious situation led Buringh (1969) to conclude that the soils probably are not, in fact, sodic.

Salt affected soils are common in Tunisia and Algeria where both surface and well irrigation waters frequently are saline. Salinization is important in other irrigated areas but on a small scale.

Magnitude of Desertification

About 25 percent of the arid region of Africa is severely desertified, with most of that represented by grazing lands and rainfed cropping lands on the south side of the Sahara (Table 6.4). The other large area that is severely affected is the mountain slopes and the plains of North Africa. Moderate to high salinity affects about 30 percent and waterlogging about 93 percent in Egypt (Aboukhaled et al, 1975). Several tens of thousands of hectares of irrigated land in Algeria and Tunisia suffer from salinization and/or waterlogging. Table 6.5 gives figures on the extent of desertification, by major land use, for African countries.

TABLE 6.4. Desertification of arid lands of Africa

Desertification class	Area affected	
	Square kilometers	Percent of arid lands
Slight	12,430,000	71.7
Moderate	1,870,000	10.8
Severe	3,030,000	17.5
Total	17,330,000	100.0

Wind erosion is dominant in the drier regions and water erosion on the wetter sloping lands. Ethiopia, Kenya, and the Maghreb countries have been subjected to especially serious water erosion, whereas wind erosion has been most damaging in sub-Saharan West Africa. While good data on the effect of land degradation on crop and livestock yields are not available, it seems likely that soil fertility losses, alone, have reduced dryland crop yields by 25 to 50 percent in the severely desertified areas. Animal productivity may well have declined by at least 50 percent nearly everywhere that domestic livestock are raised. In many areas south of the Sahara rangeland

TABLE 6.5. Arid lands affected by desertification in Africa, by countries

Country	Irrigated Land (000ha)		Rainfed Cropland (000ha)		Rangeland (000ha)	
	Total	Area affected by desertification	Total	Area affected by desertification	Total	Area affected by desertification
Algeria	292	65	4,500	4,000	83,000	76,500
Botswana	2	0.2	30	20	50,000	10,000
Cape Verde	1	0.03	45	30	0	0
Chad	3.5	0.17	1,800	1,700	97,000	96,000
Djibouti	0	0	0	0	2.2	2.2
Egypt	2,846	735	5	1	10,000	9,700
Ethiopia	30	5	3,500	3,100	85,125	77,000
Gambia	29.5	5	200	100	0	0
Kenya	20	1.1	300	270	22,000	21,000
Libya	154	12	2,500	2,000	35,200	33,600
Madagascar	670	45	200	150	5,000	4,800
Mali	120	12	2,000	1,500	108,000	106,000
Mauritania	28	0.15	150	100	72,300	71,000
Morocco	630	125	7,000	5,600	28,000	27,200
Namibia	8	0.1	10	5	66,000	16,500
Niger	16.5	0.83	4,000	3,000	104,000	101,000
Nigeria	13	0.6	5,500	5,200	30,000	28,000
Senegal	144	0.10	2,400	2,000	13,000	12,100
Somalia	100	9	1,000	950	63,600	57,500

TABLE 6.5 (continued)

Country	Irrigated Land (000ha)		Rainfed Cropland (000ha)		Rangeland (000ha)	
	Total	Area affected by desertification	Total	Area affected by desertification	Total	Area affected by desertification
South Africa	860	46	1,000	650	45,000	38,000
Sudan	1,610	250	3,500	2,600	203,000	198,000
Tanzania	40	4	2,400	1,900	28,000	14,000
Tunisia	128	50	3,000	2,100	10,100	9,000
Uganda	4	0.2	0	0	375	350
United Republic of Cameroon	2	0.3	8	7	10	6
Upper Volta	5.0	0	2,700	2,500	16,000	15,500
Zimbabwe	0	0	300	150	7,500	4,000
Total	7,756.5	1,366.78	48,048	39,633	1,182,212.2	1,026,758.2

forage production probably is less than 25 percent of the potential.

ASIA

Desertification in the arid regions of Asia is characterized by overgrazing of the rangelands of the Middle East and Central Asia, water erosion of cultivated lands from eastern China to the Mediterranean Sea, and salinization and waterlogging on a large scale in Iraq, Pakistan, China, and the Soviet Union. Mining (including oil and gas production) has caused severe damage wherever it is carried on.

Overgrazing, soil erosion, and salt damage to irrigated land are long standing problems in the Middle East and Central Asia, as is water erosion on the rainfed cultivated lands of India, Pakistan, and the loessial plateau of China. Waterlogging and salinization are centuries old problems in the lower plain of the Yellow River in China but are of relatively recent origin in the Indus Basin of Pakistan and India. Ameliorative measures to reduce land degradation have been undertaken in every country affected, with varying degrees of success. The problem is an immense one in terms of the area involved and the amount of degradation that has occurred (Figure 6.2).

Desertification Characteristics

Man first placed his imprint on the landscapes of Asia thousands of years ago. Historical records make it clear that the destructive impact of man was already

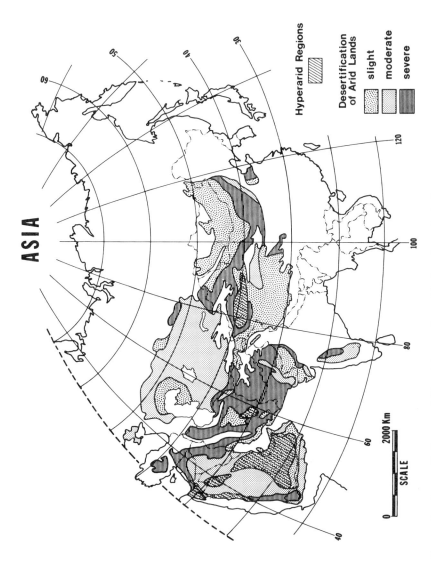

Figure 6.2. Status of desertification of arid lands of Asia.

apparent two or three thousand years ago in China and the Middle East. It was then what has been called the first period of accelerated, man-made erosion began (Dregne, 1982). Woodcutting and cultivation of sloping lands were responsible for most of the land degradation that occurred in the uplands of China and around the Fertile Crescent. Ryan (1982) believes that water erosion in Lebanon is more severe now than ever before. Sedimentation of canals, waterlogging, and salinization combined to raise havoc in the Tigris and Euphrates valleys of Mesopotamia (Jacobsen and Adams, 1958).

Grazing pressure seems to have become a much greater problem in the past two or three decades than it was before. There has been an explosion of human and livestock population recently from India (Office of Environmental Planning and Coordination, 1977) to the Middle East (Pearse, 1971). Cultivation has been extended into pastoral lands, wells have been constructed to make pastures usable throughout the year, and improved transportation facilities have made formerly remote areas accessible to livestock. As a consequence, cropping has become a high risk land use in the climatically marginal areas that formerly were the best grazing lands. At the same time, overgrazing of the remaining range lands is accelerated by the increased concentration of livestock on drier lands. Destruction of vegetative cover on sandy lands has led to widespread wind erosion, depletion of soil fertility, and the undertaking of expensive erosion control projects in China, Iran, Saudi Arabia, and elsewhere.

Magnitude of Desertification

Asia has large areas where desertification is so severe that it borders on being classed as very severe desertification, but it also has about half of the land only slightly desertified (Table 6.6). The magnitude of desertification, by major land uses, in Asian countries is shown in Table 6.7.

TABLE 6.6. Desertification of arid lands of Asia

Desertification class	Area affected	
	Square kilometers	Percent of arid lands
Slight	7,980,000	50.9
Moderate	4,480,000	28.6
Severe	3,210,000	20.5
Total	15,670,000	100.0

Water erosion is spectacular in the loessial plateau in the middle reaches of the Yellow River. The eroded silt is sometimes so thick in the river that at Shanhsien, east of Xian, the sediment load has been measured at 46 percent. The river annually carried 1.3 billion tons of sediment, enough to cover 1,800 square kilometers of land 30 centimeters deep (De Crespigny, 1971). Virtually all of that represents loess eroded from 300,000 square kilometers of highlands. Efforts are being made to reduce erosion but the problem is so huge that progress is slow. Only the great thickness of the aeolian deposits has

TABLE 6.7. Arid lands affected by desertification in Asia, by countries

Country	Irrigated Land (000ha)		Rainfed Cropland (000ha)		Rangeland (000ha)	
	Total	Area affected by desertification	Total	Area affected by desertification	Total	Area affected by desertification
Afghanistan	2,900	600	5,000	4,500	39,000	35,000
Bahrain	1	0	0	0	65	65
China	19,000	3,700	8,000	7,400	370,000	307,000
Democratic Republic of Yemen	5	0.5	0	0	28,500	22,800
India	30,000	4,500	40,000	34,000	18,000	15,000
Iran	5,250	1,320	6,000	5,500	127,000	124,000
Iraq	4,000	2,400	5,000	4,500	33,000	31,000
Israel	170	17	190	185	1,240	1,100
Jordan	60	9	1,000	950	8,400	8,200
Kuwait	1	0	0	0	1,580	1,500
Mongolia	30	4.5	900	150	93,000	56,000
Oman	35	15	0	0	21,238	20,000
Pakistan	13,300	5,050	6,000	5,400	56,000	54,000
Qatar	0.12	0	0	0	2,070	2,000
Saudi Arabia	130	26	900	800	194,000	175,000
Syria	600	150	5,000	4,000	13,000	12,000
Turkey	2,000	600	9,000	6,000	11,000	9,400
USSR	12,000	2,160	25,000	17,500	230,000	190,000

TABLE 6.7 (continued)

Country	Irrigated Land (000ha)		Rainfed Cropland (000ha)		Rangeland (000ha)	
	Total	Area affected by desertification	Total	Area affected by desertification	Total	Area affected by desertification
United Arab Emirate	5	0	0	0	8,366	7,900
Yemen Arab Republic	100	20	600	350	18,300	17,000
Total	89,587.12	20,572.0	112,590	91,235	1,273,759	1,088,965

allowed cultivation to be continued on the eroding slopes.

Water erosion approaches the spectacular also on the slopes of the dryland farming areas of western India and northern Pakistan. Fortunately, the soils there are more resistant to displacement than is loessial material. Nevertheless, the damage - in the form of reduced yields - is serious nearly everywhere, and it is getting worse.

Salinization and waterlogging are as bad - and as difficult to control - in the irrigated region of southern Iraq as they are anywhere in the world. Deep fine textured soils and almost flat topography make it hard to provide drainage that is adequate to maintain a deep water table and prevent salt accumulation. Techniques of soil and water management that are suitable for conditions in Iraq have been developed and tested successfully but the technical problems are less difficult to resolve than the social ones (Dougrameji and Clor, 1977). Land reclamation is much easier on the more permeable soils of the Indus Valley and Central Asia.

Range productivity over most of the Middle East and western Pakistan is among the lowest in the arid world. A small beginning has been made in bringing the pastoral lands up to their potential productivity. Much remains to be done, however, because range management is only a young science in Asia. Recent emphasis on sand dune stabilization has led to a number of successful efforts to find effective methods and to identify adapted plants. Accomplishments have been outstanding, under very difficult circumstances.

A rapidly growing population and limited land

resources mean that combatting desertification will be
difficult for the developing nations of Asia and very
difficult for the poor countries. Land pressures must
somehow be relieved, and that can only be accomplished
by raising productivity and yields per unit area. A
start can be made with presently available knowledge but
research and training will be needed to broaden the
knowledge base and to provide guidance to the pastoralists
and cultivators.

AUSTRALIA

Land degradation in Australia is, from the standpoint
of area affected, a problem of overgrazing of range lands
(Figure 6.3). Wind and water erosion and salinization of
irrigated and non-irrigated land are serious locally,
primarily in small areas across the southern part of the
continent, and their economic impact probably exceeds
that due to overgrazing. Land degradation is usual
around communities and popular tourist areas such as
Ayers Rock.

Overgrazing began 50 to 100 years ago with settlement
of the interior arid lands at a time when there was no
information on the carrying capacity of range land and
little knowledge of climatic averages and extremes.
During the first quarter of the 20th century, problems of
wind and water erosion, salinization and waterlogging
of irrigated land, and saline seepage of dryland areas
arose as cultivation spread into the drier areas.
Remedial action was undertaken on a concerted basis in
the 1930s and 1940s when states enacted soil conservation
legislation and established units to correct land abuses

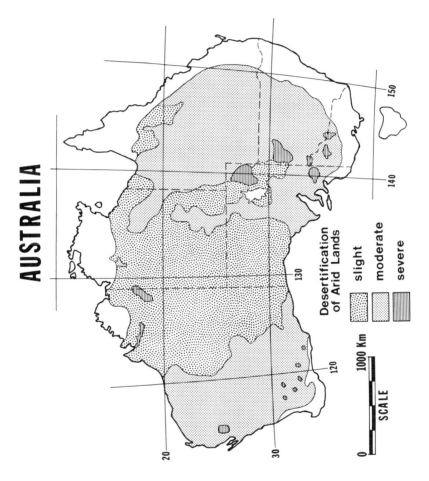

Figure. 6.3. Status of desertification of arid lands of Australia.

and prevent further deterioration. At present, it
probably is safe to say that overgrazing is much less of
a problem than it was before 1940, that wind and water
erosion still occur but to a lesser degree than before,
that salinization of irrigated land in the Murray River
catchment (watershed) is increasing, and that saline
seepage of semi-arid and subhumid lands is expanding
steadily.

Desertification Characteristics

Australia has about 25 percent of its arid region
unoccupied (Condon, 1978) due to the unpalatability of
spinifex grasses or to the remoteness of the land from
population centers. Most of the remaining 75 percent
is moderately desertified. Severe desertification is
most extensive in the saltbush-bluebush (Atriplex-
Maireana) vegetation type occurring in New South Wales and
South Australia, where overgrazing has caused degeneration
of the plant cover. Degradation of the vegetation due to
overgrazing has also been severe on the fine texture
lowland soils (Vertisols) of eastern Australia and in
the flood plains and surrounding slopes of coastal river
valleys, particularly the Gascoyne, Ord, and Victoria
catchments in Western Australia and the Northern
Territory. Gully erosion has been severe on the Ord
River catchment and significant on the other catchments,
with greatly increased runoff where the trampling of
livestock has developed water-repellent surface crusts.
The presence of large numbers of feral donkeys in the
Victoria River watershed compounds the problem of grazing

control there.

Areas delineated on the map as severely desertified are not uniformly degraded. The typical spotty character of desertification shows up in river catchments as badly overgrazed and gullied land on flood plains and on the gentle slopes adjoining them, with moderately degraded land on the rocky slopes and those parts of the catchment that are farther from watering points. Degradation of vegetation is more uniform in the saltbush-bluebush zones of southeastern Australia but there are many slightly and moderately desertified areas in the less accessible parts of those shrub steppe zones.

Distribution of saline seeps on drylands, and salinized and waterlogged irrigated lands (Mabbutt, 1978), is also spotty. Seepage of salt-laden water on slopes has plagued cultivated lands in the semi-arid and subhumid regions of Western Australia, South Australia, New South Wales, and Victoria for decades. The resulting saline soils on lower slopes and in waterlogged depressions ordinarily cover a few or several hectares and are surrounded by non-saline soils. In Western Australia where the problem is most acute, seepage of saline groundwater into streams has also had a significant adverse effect on water quality.

Waterlogging in the Murray River and its tributaries is characterized by groundwater mounds affecting tens of thousands of hectares of irrigated land (Pels, 1978). The high groundwater levels lead to secondary salinization of soils as the salts originally present in the subsoils are carried to the surface through evaporation from the shallow water tables. Since waterlogging and

salinization are functions of the irrigation water
management and the presence or absence of a drainage
system, salty land is interspersed with non-salty land.

Wind and water erosion have been severe in the past
but are less so now, by and large. Wind erosion is an old
problem in the grazing and cultivated lands of New South
Wales (Wagner, 1978) and the wheat lands of the Murray
Mallee (Matheson, 1978). One of the special problems
in semi-arid Australia is the appearance of "scalds"
when sandy surface soils are blown or washed away and
the slowly permeable subsoil of texture-contrast (duplex)
soils is exposed. Revegetating scalds is difficult
unless steps are taken to impound water on the surface.
Mining towns such as Broken Hill have experienced
accelerated wind erosion and dust storms as the land around
the towns was denuded of vegetation.

Magnitude of Desertification

A 1969 report on land degradation in Australia
(Newman and Condon, 1969) included an estimate that about
14 percent of the arid rangelands were severely degenerated
in pastoral value, 23 percent were moderately degenerated,
and 63 percent showed minor or little or no degradation.
Condon (1978) notes that much of the serious erosion and
degeneration of rangelands which was evident until the
late 1940s was a consequence of heavy use during the
late 1800s. Since the 1940s there has been considerable
improvement in range conditions, partly due to abnormally
favorable rainfall over the last thirty years. The
magnitude of desertification in Australia is shown in

Table 6.8 and Table 6.9.

TABLE 6.8. Desertification of arid lands of Australia

Desertification class	Area affected	
	Square kilometers	Percent of arid lands
Slight	2,330,000	44.2
Moderate	3,510,000	47.6
Severe	520,000	8.2
Total	6,360,000	100.0

About 1 percent of the cultivated land in Western Australia is affected by saline seepage, with lesser percentages affected in the other states. Northcote and Skene (1972) estimate that 85,000 hectares of irrigated land in Australia are salt-affected, most of it in the Murray Basin. They also state that at least 197,000 hectares of dryland are affected by secondary salinization. Waterlogging and salinization go hand-in-hand in the Murray region.

Techniques to control desertification are well-known in Australia, the problem is recognized as a serious one, and state agencies are available to assist in carrying out reclamation and prevention programs. Grazing pressure probably has been reduced by half since the late 1800s and early 1900s and wind and water erosion in the more populous southern part of the continent is much less of a problem now. Soil salinity and waterlogging are more difficult to control because of the cost arising from

TABLE 6.9. Arid lands affected by desertification in Australia

	Irrigated Land (000ha)		Rainfed Cropland (000ha)		Rangeland (000ha)	
	Total	Area affected by desertification	Total	Area affected by desertification	Total	Area affected by desertification
Country Australia	1,600	160	2,000	1,500	550,000	307,000

the special conditions in the Murray irrigated area. One method used to dispose of saline drainage water without increasing the salinity of Murray River water downstream is to divert it into evaporation basins. The principle is good but the cost is high and there are uncertainties about future contamination of groundwater by effluent from the basins.

NORTH AMERICA

Overgrazing has left a lasting imprint on the arid lands of North America, and little has been accomplished in attempts to restore the land to its original productivity. Wind and water erosion are extensive and salinization and waterlogging - of varying degrees of severity - are common in nearly all of the irrigated valleys. At present, desertification probably has been stabilized, with some improvement in plant cover of rangelands and in control of erosion and waterlogging over the past 30 years. Salinization of irrigated land is still occurring and the dryland saline seep problem is getting worse.

The overgrazing that destroyed or severely altered the original grass cover of the rangelands of Mexico and the United States began in Mexico after the Spanish conquest and spread into the U.S. Southwest (Figure 6.4). By the early part of the 19th century, overgrazing was already a fact on both sides of the international boundary. With the explosive expansion of cattle numbers in the Southwest when the railroads arrived, range carrying capacities were greatly exceeded, and that situation continued well into the 20th century. The increased gully

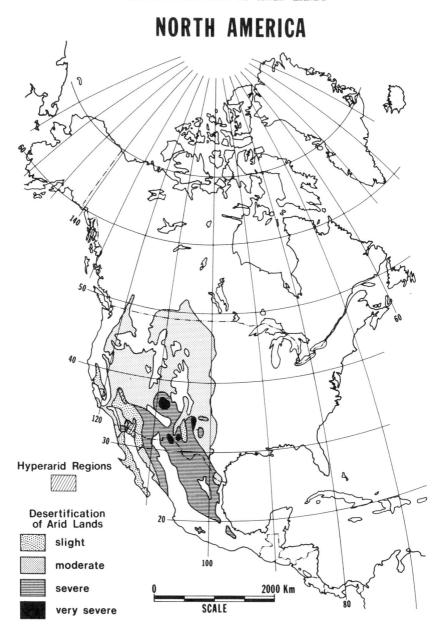

Figure 6.4. Status of desertification of arid lands of North America

(arroyo) formation that occurred in overgrazed rangelands during the latter part of the 19th century has been associated with range deterioration but it is not clear that overgrazing was the cause (Cooke and Reeves, 1976).

Range deterioration, erosion, and salinization and waterlogging first received attention from research organizations in the late 1800s and early 1900s. Many of the basic principles for controlling desertification were established at that time. However, application of the principles did not become widespread until the 1940s and 1950s. Even though solutions are available now, much remains to be done to put them into practice.

Desertification Characteristics

There are about 450 million hectares in the arid regions of Canada, the United States, and Mexico. Approximately two-thirds of that total is moderately desertified and less than one-third severely desertified, with a considerable area of slightly desertified land and four small delineations of very severely desertified land.

The slightly desertified area encompasses the Mojave Desert and the Sonoran Desert in the states of California, Arizona, Sonora, and Baja California. The climate is too dry and the topography too rugged for significant overgrazing to have occurred. Salinity problems do occur in some of the irrigated valleys along the Colorado and Gila rivers but the areas are too small to be shown on the map.

In general, rangelands are moderately desertified in

the north and severely desertified in the south. The
dividing line approximates the ecotone between sagebrush
(cool climate) and creosotebush (hot climate).
Desertification is the result of a combination of overgrazing and water erosion. The four delineations of very
severely desertified land on the map are in rangelands.
Three of those four areas in New Mexico, Texas, and
Chihuahua suffered first from overgrazing, then from wind
erosion. The result is small hummocks of mesquite-topped
sand dunes separated by completely barren blow-outs.
The fourth - and largest - of the very severely desertified
areas is on the Navajo Indian Reservation of northern New
Mexico and Arizona. Overgrazing by sheep has removed
most of the perennial grasses and exposed the easily
eroded shale-derived soils to extensive sheet and gully
erosion. It is unlikely that any of these four areas
will recover naturally in the next hundred or more years
even if grazing is completely prohibited.

Overgrazing in the U.S. Southwest is not a recent
condition. The number of cattle in the Arizona Territory
(the states of Arizona and New Mexico) rose from 5000 in
1870 to 1,095,000 in 1890, a phenomenal 20,000 percent
increase. By 1901 the rangeland in Arizona was said to
be seriously overgrazed and by 1908 the range in New
Mexico was deteriorating (Herbel, 1979). From 1866 to
1934, lack of grazing control on federal rangelands
resulted in extensive overgrazing, an example of the
"tragedy of the commons". Extending cultivation into
climatically marginal zones destroyed forage and subjected
the land to severe wind and water erosion. The destruction
sometimes was so severe that it hampered revegetation

attempts that were made after the land was abandoned. Noxious brush density also increased rapidly after 1900. In Texas, 82 percent of the grasslands are infested with low-value woody plants, and 50 percent are covered with dense stands of brush (Herbel, 1979). The brush will have to be controlled or eradicated before improved management or seeding or water-spreading can be effective. After sagebrush (Artemesia spp.) land was converted to perennial grass in western Colorado, runoff did not change appreciably but sediment yield (erosion) was greatly reduced (Hadley, 1977).

Mexican rangelands probably were managed fairly well during the period from 1880 to 1910, thanks to government policies, but overgrazing and cereal cultivation in the dry regions have caused significant land degradation (Medellin Leal and Gomez Gonzales, 1979).

Wind erosion on dry farmed lands is the second most extensive desertification process. The ravages of the drought-inspired Dust Bowl of the 1930s were but an extreme example of the damage wind erosion has caused in the western Great Plains, the Pacific Northwest and elsewhere. Wind erosion of cultivated land, whether dry farmed or irrigated, has been a problem ever since the original sod cover was broken in the dry regions. It is an ever-present threat, worsening during droughts and lessening during wet periods, but always occurring somewhere every year. While sandy soils are the most susceptible, even fine textured soils will blow when conditions are right, and along with the dust goes organic matter and nutrient elements. The introduction of sprinkler irrigation systems, especially the center pivot systems, has enabled previously unsuitable rolling sandy lands to be cropped successfully. If and when those soils are abandoned, for reasons of economy or shortage of water, the United States will face an even greater wind erosion threat than it has had to cope with in the past.

Most of the Great Plains crop lands are moderately desertified by wind erosion. Three severely desertified areas in the southern plains have been delineated. They represent sandy soils which have been adversely affected by wind erosion.

Accelerated water erosion has been especially serious on overgrazed rangelands and on crop land in the Palouse region of the Pacific Northwest and in the highlands of Mexico. Normal geologic erosion has been severe in the Badlands of South Dakota and in the shale materials of Wyoming, Colorado, Utah, and New Mexico but man has managed to make matters worse through overgrazing and cultivation.

While salinity and waterlogging problems are found in nearly every irrigated valley, the only areas large enough to be shown are in the San Joaquin Valley of California, the contiguous Imperial and Mexicali valleys, and the lower valley of the Rio Grande. Land degradation is more severe in the Mexicali Valley than in the adjoining Imperial Valley because 1) the salinity of the irrigation water is worse where Mexico diverts Colorado River water than where the Imperial Valley does and 2) the Mexicali Valley does not have the intensive drainage network that the Imperial Valley has. Waterlogging is more of a problem in the northern states of the United States, whereas the combination of waterlogging and salinization is more serious in the south. Serious salinity problems are associated with marine shales in the numerous small irrigated valleys in the upper Colorado River watershed.

The special problem of saline seeps in non-irrigated crop lands is widespread in the states of Montana, South Dakota, and North Dakota and in the prairie provinces of Manitoba, Saskatchewan, and Alberta (Vander Pluym, 1978). The affected areas are too scattered to be delineated on the map.

Soil compaction and soil crusting are problems associated with coarse, medium, and fine textured soils. Man-induced compacted subsoil layers are common in fine textured irrigated soils, especially, and in medium to coarse textured dryland soils. Surface crusting occurs on irrigated, dryland, and rangeland soils. Effects on crop yields are greatest in irrigated coarse and medium textured soils.

Magnitude of Desertification

Nearly 90 percent of North American arid lands are moderately and severely desertified (Table 6.10). Slightly more than 80 percent of the rangelands in the western United States are estimated to be in fair or poor condition. That figure is not significantly different from a 1936 estimate of the condition of federal rangelands, only. What has changed greatly is the amount of land in poor condition. It was 58 percent in 1936 and 32 percent in 1972. There has been little change since 1966. The 1936 figure probably reflects the short-term effect of the drought of the 1930s rather than a long-term condition. In any event, the status of the U.S. rangelands does not appear to have been much improved in recent years, and less than 20 percent is producing anywhere near its potential. It is likely that the same condition prevails in Mexico. Table 6.11 gives figures on the extent of desertification, by major land uses, in North America.

TABLE 6.10. Desertification of arid lands of North America

Desertification class	Area affected	
	Square kilometers	Percent of arid lands
Slight	440,000	9.8
Moderate	2,720,000	61.6
Severe	1,200,000	27.1
Very severe	67,000	1.5
Total	4,427,000	100.0

Data on the status of wind and water erosion in the arid regions are not available. A 1979 report of the U.S. Soil Conservation Service said that 58 percent of the nation's cropland needed conservation treatment of an unspecified character. That figure was down from 64 percent in 1967. The major conservation treatment undoubtedly would be water erosion control.

An indication of the magnitude of the wind erosion problem can be obtained from annual surveys conducted by

TABLE 6.11. Arid lands affected by desertification in North America, by countries

	Irrigated Land (000ha)		Rainfed Cropland (000ha)		Rangeland (000ha)	
	Total	Area affected by desertification	Total	Area affected by desertification	Total	Area affected by desertification
Country						
Canada	300	60	5,000	3,000	10,000	7,000
Mexico	3,750	1,125	7,500	6,700	100,000	96,000
USA	15,500	1,650	30,000	15,000	235,000	188,000
Total	19,550	2,835	42,500	24,700	345,000	291,000

the U.S. Soil Conservation Service for the Great Plains. Since 1935, surveys have been made of the amount of land damaged by wind erosion in susceptible parts of the Great Plains. Annual totals have varied from about 400,000 hectares to a maximum of 6,000,000 hectares. The lowest amount of land damage occurred during the wet 1940s and 1960s; the highest was in 1955 during a protracted drought. In any one year, the percentage of damaged cropland varied from 0.6 to 9 percent of the total cropland surveyed. There is no discernible trend upward or downward in the area damaged; the dryness of the particular year seems to determine how bad wind erosion will be.

Wind erosion in the prairie provinces of Canada - and the associated loss in soil fertility - has long been recognized as a serious problem (Soil Research Laboratory Staff, 1949). Soil and crop conditions are similar on both sides of the international boundary in the northern Great Plains, and so are erosion problems. Wind erosion is much less of a problem than water erosion in the arid regions of Mexico, where sandy soils are less prevalent in cropped areas than is the case in Canada and the United States.

Salinization is estimated to affect adversely about 11 percent of the irrigated land in the United States and about 30 percent in Mexico. Dryland saline seepage is a Canada-United States problem, with about 0.8 million hectares severely affected and perhaps another 2.7 million slightly or moderately affected. Waterlogging may affect about the same area as salinization but the two problems are not always found together.

Techniques for controlling desertification in North America are well known. Recognition of the seriousness of the problem varies from year to year, usually depending upon the amount of notoriety droughts or floods receive. In the United States, where soil conservation has been given considerable financial support at the federal level, only about one-third of the nation's cropland has been adequately treated against soil erosion. Much less than one-third of the depleted rangeland has been improved. The situation may be better in Canada but probably is worse in Mexico.

The reasons for the relatively poor performance in combatting desertification are numerous. High among them is the emphasis on short-term versus long-term benefits, inability to finance improvements, lack of clear-cut proof that control is profitable in the immediate future, and the absence of a public sense of urgency toward controlling land degradation.

SOUTH AMERICA

Overgrazing, cultivation of unsuitable land, and land degradation by woodcutting - and the associated water erosion - are long-standing problems in the arid regions of South America (Figure 6.5). Wind erosion is a major threat in the semi-arid pampas in Argentina, where moving sand dunes have ruined much land. Salinity and waterlogging have adversely affected irrigated regions in western Argentina, particularly along the Rio Salado, and in the numerous narrow irrigated valleys crossing the coastal plain of Peru. The very large salinas and salares (salt lakes) found in western Argentina, northern

210 DESERTIFICATION OF ARID LANDS

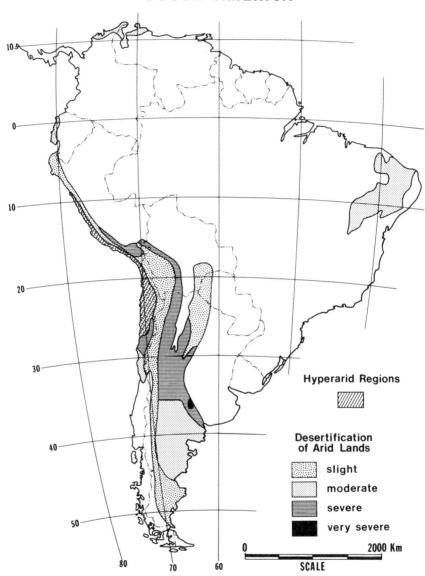

Figure 6.5. Status of desertification of arid lands of South America

Chile, and the altiplano of Bolivia and Chile are of natural origin rather than being due to man's activities. There is little evidence of land improvement, and conditions are either worsening or at least not improving nearly everywhere.

Desertification Characteristics

Heavy woodcutting for mine timbers and fuel accompanied the settling of western South America by the Spaniards in the 16th and 17th centuries. Along with that went localized overgrazing of livestock and the cultivation of the nearby slopes, with the inevitable accelerated erosion. It was not until the 19th and 20th centuries, however, that land degradation on the slopes of the Andes Mountains and the coastal ranges became widespread. Population increases and land tenure systems in effect in most countries of South America have combined to put further pressure on the arid regions in recent years.

Desertification has been held to a moderate level in semi-arid northeastern Brazil by the erratic character of the rainfall there. Land destruction probably would have been much worse if the rainfall were more reliable and if severe droughts did not occur so frequently. The original caatinga vegetation is xerophytic and is adapted to the long dry periods that occur each year (Banco do Nordeste do Brasil, 1964).

The coast of Peru is crossed by a large number of short rivers flowing from the Andes to the Pacific Ocean. Many of the irrigated valleys are affected to some degree by salinization and waterlogging. The valleys constitute only a small part of the coastal desert of Peru and even less of the Chilean desert, however. Most of the desert has experienced little development and little desertification.

Farther south, in the semi-arid coastal mountains of Chile, land degradation due to overgrazing and cultivation of sloping lands has been severe around population centers. Indiscriminate woodcutting has also been an important negative factor in the development of the region.

In Argentina, which has more arid land than any other South American country, overgrazing has led to the degradation of range vegetation from the high plateaus in the north to the cold Patagonian desert in the south. Wind erosion plagues both range and cultivated lands, especially in the southern half of the nation.

Magnitude of Desertification

Desertification affects about 56 percent of the arid lands of South America to a moderate or worse degree (Table 6.12) and is a common problem in the rainfed cropping and rangelands (Table 6.13). Overgrazing of rangelands has been the most destructive aspect of desertification in all the arid regions of South America. There has been a progressive elimination of palatable species by cattle, sheep, and goats over many decades. Along with this has gone the cutting of trees for timber, firewood, and charcoal. In the Gran Chaco of Argentina it is difficult to find good stands of the valuable red and white quebracho trees that once grew everywhere.

TABLE 6.12. Desertification of arid lands of South America

Desertification class	Area affected	
	Square kilometers	Percent of arid lands
Slight	1,340,000	43.6
Moderate	1,050,000	34.1
Severe	680,000	22.1
Very severe	6,000	0.2
Total	3,076,000	100.0

Water erosion has followed overgrazing throughout the continent, and wind erosion has become an acute problem in parts of Argentina (Prego et al, 1971). The most serious wind damage has been done in the dry farmed areas of central Argentina where cultivation has led to the formation of mobile dunes. In 1963 an estimated 16,000,000 hectares of land were affected to varying degrees by wind erosion (Comité Argentino, 1963), and the situation has become worse since then. Soil fertility has declined and runoff has increased.

Salinization and waterlogging affect about 20 percent of the irrigated land in the coastal valleys of Peru (Comité Peruano de Zonas Aridas, 1963) and large sections of the semi-arid Gran Chaco in and around the province of Santiago del Estero. The same problem, to a lesser degree, appears in the irrigated regions in arid western Argentina and in northeastern Brazil.

Although only about 22 percent of the arid regions of

TABLE 6.13. Arid lands affected by desertification in South America, by countries

Country	Irrigated Land (000ha)		Rainfed Cropland (000ha)		Rangeland (000ha)	
	Total	Area affected by desertification	Total	Area affected by desertification	Total	Area affected by desertification
Argentina	1,500	310	5,000	3,800	180,000	126,000
Bolivia	65	6	1,000	950	12,000	11,500
Brazil	520	78	6,000	5,000	140,000	135,000
Chile	1,280	320	1,400	1,350	24,000	22,400
Colombia	0	0	0	0	3,500	3,200
Ecuador	460	115	40	39	300	280
Paraguay	9	2	50	20	12,000	9,600
Peru	1,155	346	500	450	9,500	8,800
Venezuela	350	52	300	250	2,800	2,600
Total	5,389	1,229	14,290	11,859	384,100	319,380

South America are severely or very severely affected by desertification (Table 6.12), the problem is critical because the best lands have been degraded and the degradation is continuing. Virtually no improvement of grazing lands has occurred anywhere, nor have soil salinization and waterlogging been controlled. Some progress has been made in stabilizing sand dunes in the dry farming areas of Argentina but much remains to be done. Water erosion in the Andes and the semi-arid western coastal ranges continues to present problems and may be getting worse. A fast-growing population virtually assures greater desertification in the years ahead unless governments become determined to cope with the problem.

SPAIN

All of the arid regions of Spain have been moderately to severely desertified for decades, if not for centuries (Figure 6.6). Most of the damage has been done by heavy grazing and woodcutting but wind and water erosion on cultivated land has also been extensive. Salinization and waterlogging of irrigated land is a major problem in the Guadalquivir Valley in the south and is a locally important problem in parts of the Ebro River watershed in northeast Spain. Destruction of the native vegetation and the subsequent water erosion of thin soils on slopes has had a devastating effect on the plant environment and on the potential productivity of the land. Mobile sand dunes are found along the Mediterranean coast.

216   DESERTIFICATION OF ARID LANDS

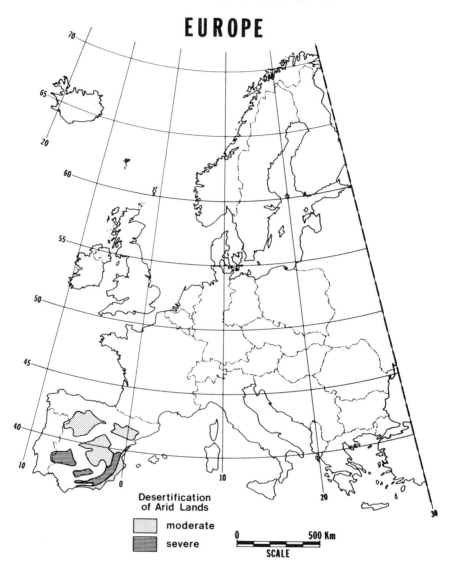

Figure 6.6.   Status of desertification of arid lands of Spain

Desertification Characteristics

During the past several centuries, heavy grazing by sheep and goats has led to the destruction of much of the herbaceous and woody vegetation on the non-cultivated land (Albareda, 1955). Water erosion has been severe on the overgrazed slopes as a result of the loss of vegetative cover and the torrential character of the rains. Cutting of wood for fuel and construction and the extension of dryland farming into the pasture lands has accompanied overgrazing. Plant cover has changed to a more xeric type and surface runoff has increased.

A monoculture of grain in the cultivated regions has depleted the native fertility of the soil and has been responsible for increasing the susceptibility of the land to wind and water erosion. Extended droughts from time to time have served to accelerate desertification. Water erosion is severe nearly everywhere on sloping land.

Salinization and waterlogging do not affect a high percentage of the total cultivated land in Spain but important and large areas of affected soils do occur in irrigated valleys. The major salt affected areas in the northeast are in the Ebro River watershed in the vicinity of Zaragosa and Herida. Seepage water from irrigation on the higher land has caused waterlogging and salinization of lower lying areas (Martínez, 1978). Gypsum is a common constituent of the soils.

The other major salt affected areas are in southwest Spain near the coast. The soils are composed of fine textured sediments that were subjected regularly to flooding by sea water in the past. Surface and internal drainage of the irrigated land is poor and water tables generally are close to the surface. Pumping is required to lower the water tables (Ayers et al, 1960). Irrigation water usually is low in salts but some is quite saline, especially in southeast Spain.

Magnitude of Desertification

Approximately 50 percent of Spain is arid. In the arid regions about 60 percent is moderately desertified and 40 percent is severely desertified (Table 6.14). Virtually all of the rangeland has been degraded (Table 6.15), and most of it has suffered severe land degradation. Range productivity is probably stabilized at a low level now, with little improvement anywhere. Erosion continues on the extensive dryfarm lands except in a few places where soil and water conservation measures have been instituted. Soil fertility remains low.

TABLE 6.14. Desertification of arid lands of Spain

Desertification class	Area affected	
	Square kilometers	Percent of arid lands
Slight	0	0
Moderate	140,000	70.7
Severe	60,000	29.3
Total	200,000	100.0

TABLE 6.15. Arid Lands affected by desertification in Spain

	Irrigated Land (000ha)		Rainfed Cropland (000ha)		Rangeland (000ha)	
	Total	Area affected by desertification	Total	Area affected by desertification	Total	Area affected by desertification
Country						
Spain	2,400	890	5,000	4,200	16,000	15,500

Salinization and waterlogging affect some 240,000 hectares of irrigated land to various degrees (Ayers et al, 1960). Reclamation has been undertaken in several areas. Because of the need for more agricultural production and the gradual worsening of the salt problem, irrigation has received special attention in recent decades. Much remains to be done.

Soil and water conservation techniques for dryland farming are known but their application to the field is limited. Range management is not a well supported science in Spain. Considerable progress has been made in the reclamation of saline soils and procedures for doing so are quite well known (Martínez, 1978).

CONCLUSIONS

Man has been a poor steward of his environment in the arid regions of the world. Only in places where the environment is too inhospitable for large scale development, such as in the Sahara or the polar deserts, has the imprint of man been insignificant. For the 19 percent of land that is severely or very severely desertified, the degradation that has occurred has made a permanent change in the soil resource. There is no possibility of restoring the land to its original, pre-man condition in the foreseeable future. The same can be said about a part of the 29 percent that is moderately desertified. With good management, the productivity of virtually all of the arid lands capable of supporting a fair amount of vegetative cover can be raised significantly, despite the soil degradation. Only the very severely desertified land - a very small part of the total - is beyond saving.

OCCURRENCE OF DESERTIFICATION

Desertification, principally in the form of salinization, is least extensive on irrigated land. Since salinity control techniques are well known and easily transferable and the economic returns are high, reclamation of irrigated lands should have a high priority. The situation becomes less favorable on rainfed croplands because the vagaries of weather make investments more risky. At the same time, it should be noted that the biggest anthropogenic problem on rainfed croplands is the loss of fertility due to crop removal and erosion. Fertility can be restored easily, but usually expensively, by the addition of commercial fertilizers. The dryland farming soils that are the most difficult to improve are the coarse sandy soils and those on steeply sloping land. Desertification affects a greater percentage of grazing land than of irrigated or rainfed cropland, and the economic benefit of range improvement is much lower than it is for the other lands.

The current rate of desertification is high in the rangelands and cultivated lands of Mexico, Algeria, Morocco, Egypt, The Sudan, Ethiopia, Kenya, Somalia, Iraq, Iran, Pakistan, and China. The rate is moderate in most of the remaining arid region countries and is low in a few countries such as Australia and Saudi Arabia. A low or moderate current rate does not necessarily mean that desertification is under control. In many countries, the desertification rate is only moderate because much of the possible damage has already been done and productivity has become stabilized at a low level.

REFERENCES

Aboukhaled, A., Arar, A., Balba, A. M., Bishay, B. G., Kadry, L. T., Rijtema, P. E., and Taher, A. 1975. Research on crop water use, salt affected soils and drainage in the Arab Republic of Egypt. Near East Regional Office, FAO, Cairo. 92 p.

Agricultural Research Service, 1974. Review of ARS Research Program on Western Range Ecosystems. Technical Report, U.S. Department of Agriculture, Washington, D.C. 108 p.

Albareda, J. M. 1955. Influence des changements de la vegetation dans les sols arides. In Plant Ecology, Arid Zone Research V, Unesco, Paris, pp. 84-88.

Ayers, Alvin D., Vasquez, A., de la Rubia, J., Blasco, F., and Samplon, Sabino. 1960. Saline and sodic soils of Spain, Science 90: 133-138.

Banco do Nordeste do Brasil. 1964. O Nordeste e as Lavouras Xerofilas. Banco do Nordeste do Brasil, S.A., Departamento de Estudos Economicos do Nordeste, Fortaleza, Ceara. 238 p.

Bogges, William, McGrann, James, Boehlje, Michael, and Heady, Earl O. 1979. Farm level impacts of alternative soil loss control practices. Journal of Soil and Water Conservation 34: 117-183.

Box, Thadis W. 1977. The Arid Lands Revisited--One Hundred Years Since John Wesley Powell. Faculty Association, Utah State University, Logan, Utah. 30 p.

Buringh, P. 1969. Sodic vertisols in central Sudan. In I. Szabolcs (editor), Symposium on the Reclamation of Sodic and Soda-Saline Soils, Agrokemia es Talajtan, Tom. 18, Supplementum, pp. 100-102.

Charreau, Claude. 1974. Soils of tropical dry and dry-wet climatic areas of West Africa and their use and management. Preliminary Draft. Agronomy Mimeo 74-26, Cornell University, Ithaca, New York. 434 p.

Comité Argentino. 1963. Las Tierras Aridas y Semiaridas
de la República Argentina. Informe Nacional.
Conferencia Latinoamericana para el Estudio de las
Regiones Aridas, Buenos Aires, Argentina. 54 p.

Comité Peruano de Zonas Aridas. 1963. Informe Nacional
sobre las Zonas Aridas. República del Perú,
Ministerio de Agricultura, Lima. 105 p.

Condon, R. W. 1978. Land tenure and desertification in
Australia's arid lands. Search 9: 261-264.

Cooke, Ronald H., Reeves, Richard W. 1976. Arroyos and
Environmental Change in the American South-West.
Clarendon Press, Oxford, England. 213 p.

Dahl, Gudrun, and Hjort, Anders. 1979. Pastoral change
and the role of drought. Swedish Agency for Research
Cooperation with Developing Countries Report R2:
1979, Stockholm, Sweden. 50 p.

De Crespigny, R. R. C. 1971. China: The Land and Its
People. St. Martin's Press, New York, N.Y. 235 p.

Delwaulle, J. C. 1973. Désertification de l'Afrique au
sud du Sahara. Bois et Forêts des Tropiques,
149: 3-20.

Delwaulle, J. C. 1977. La situation forestiére dans le
Sahel. Bois et Forêts des Tropiques, 173: 3-22.

Depierre, D., and H. Gillet. 1971. Désertification de la
zone Sahelienne au Tchad. Bois et Forêts Tropiques
139: 3-25.

Dougrameji, J. S., and Clor, M. A. 1977. Case Study on
Desertification. Greater Mussayeb Project. Iraq.
United Nations Conference on Desertification A/CONF.,
74/10. 102 p.

Dregne, H. E. 1982. Historical perspective of accelerated
erosion and effect on world civilization. In
Determinants of Soil Loss Tolerance. ASA Publication
Number 45, American Society of Agronomy, Madison,
Wisconsin, pp. 1-14.

Elgabaly, M. M. 1969. Three types of sodic soils in the United Arab Republic. In I. Szabolcs (editor), Symposium on the Reclamation of Sodic and Soda-Saline Soils, Agrokemia es Talajtan, Tom. 18; Supplementum, pp. 87-89.

Floret, Christian, and Hadjed, Mohamed S. 1977. An attempt to combat desertification in Tunisia. Ambio 6: 366-368.

General Accounting Office. 1977. To protect tomorrow's food supply soil conservation needs priority attention. Report to the Congress, CED-77-30, Washington, D.C. 59 p.

Grove, A. T. 1974. Desertification in the African environment. African Affairs 73: 137-151.

Hadley, R. F. 1977. Evaluation of land-use and land-treatment practices in semi-arid western United States. Phil. Trans. Royal Soc. London, B, 278: 543-554.

Herbel, Carlton H. 1979. Utilization of grass- and shrublands of the south-western United States. In B. H. Walker (editor), Management of Semi-Arid Ecosystems. Elsevier Scientific Publishing Company, New York, pp. 161-203.

Jacobsen, Thorkild, and Adams, Robert M. 1958. Salt and silt in ancient Mesopotamian agriculture. Science 128: 1251-1258.

Mabbutt, J. A. 1978. Desertification in Australia. Water Research Foundation of Australia Report No. 54, Kingsford, N.S.W., Australia. 132 p.

Martínez Beltran, J. 1978. Drainage and reclamation of salt-affected soils in the Bardenas area, Spain. International Institute for Land Reclamation and Improvement Publication No. 24, Wageningen, The Netherlands. 321 p.

Matheson, W. E. 1978. Soil loss made South Australia come down to earth. Journal of the Soil Conservation Service of N.S.W., 34: 88-100.

Medellin Leal, F., and A. Gomez Gonzales. 1979.
Management of natural vegetation in the semi-arid
ecosystems of Mexico. In B. H. Walker (editor),
Management of Semi-Arid Ecosystems. Elsevier
Scientific Publishing Company, New York, pp. 351-376.

Newman, J. C., and Condon, R. W. 1969. Land use and
present condition. In R. O. Slater and R. A. Perry
(editors), Arid Lands of Australia, Australian
National University Press, Canberra, Australia,
pp. 105-132.

Northcote, K. H., and Skene, J. K. M. 1972. Australian
Soils with Saline and Sodic Properties. C.S.I.R.O.
Soil Publication No. 27, Melbourne, Australia. 62 p.

Office of Environmental Planning and Coordination. 1977.
Country Report: India. Department of Science and
Technology, Government of India, New Delhi. 71 p.

Pearse, C. K. 1971. Grazing in the Middle East: Past,
present, and future. Journal of Range Management
24: 13-16.

Pels, Simon. 1978. Waterlogging and salinisation in
irrigated semi-arid regions of NSW. Search 9: 273-276.

Pereira, H. C. 1977. Land-use in semi-arid southern
Africa. Phil. Trans. Royal Soc. London, B, 278:
555-563.

Prego, Antonio J., Ruggiero, Roberto A., Alberti,
Florentino Rial, and Prohaska, Federico J. 1971.
Stabilization of sand dunes in the semiarid Argentine
pampas. In William J. McGinnies, Bram J. Goldman,
and Patricia Paylore (editors), Food, Fiber, and the
Arid Lands, University of Arizona Press, Tucson,
pp. 369-392.

Rapp, Anders. 1974. A review of desertization in
Africa - Water, vegetation, and man. Secretariat for
International Ecology, SIES Report No. 1, Stockholm,
Sweden. 77 p.

Ryan, J. 1982. Soil erosion in Lebanon. Agronomy
Abstracts, American Society of Agronomy, Madison,
Wisconsin, p. 256.

Soil Research Laboratory Staff. 1949. Soil Moisture, Wind Erosion and Fertility of Some Canadian Prairie Soils. Department of Agriculture Publication 819, Ottawa, Canada. 78 p.

Vander Pluym, Hank S. A. 1978. Extent, causes, and control of dryland saline seepage in the northern Great Plains region of North America. In H. S. A. Vander Pluym (editor), Dryland-Saline-Seep Control, Agriculture Center, Lethbridge, Alberta, 1: 48-58.

Wagner, Rex. 1978. Soil conservation in New South Wales, 1938-1978. Journal of the Soil Conservation Service of N.S.W., 34: 124-132.

Weaver, J. E., and Albertson, F. W. 1940. Deterioration of midwestern ranges. Ecology 21: 216-236.

Widstrand, Carl Gosta. 1975. The rationale of nomad economy. Ambio 4: 146-153.

INDEX

Africa   14, 96, 129, 137, 155, 174-185

Albedo   96

Alberta   205

Algeria   179, 182, 221

Andes Mountains   211

Argentina   12, 112, 138, 209, 212, 213, 214

   Gran Chaco   212, 213

   Rio Salado   112, 209

   Santiago del Estero   213

Arid regions, map of   6

Arizona   201, 202

Arizona Territory   202

Asia   14, 107, 129, 155, 177, 185-192

Assessment (desertification)   84, 86, 87, 91

Aswan   117

Atacama Desert   120

Australia   14, 100-102, 121, 124, 129, 130, 153, 155, 180, 192-199, 221

   Ayers Rock   125, 153, 192

   Broken Hill   196

   Cobar   121

   Gascoyne Basin   101, 194

    Murray Basin   197

    Murray Mallee   196

    Murray River   194, 195

    New South Wales   194-196

    Northern Territory   194

    Ord (River)   194

    South Australia   194, 195

    Victoria   195

    Victoria River   194

    Western Australia   197

Ayers Rock   125, 153, 192

Baja California   201

Bolivia   211

Brazil   211, 213

Broken Hill   196

California   124, 201, 204

Canada   201, 208, 209

    Alberta   205

    Manitoba   205

    Saskatchewan   205

Chad   3

Chihuahua   202

INDEX

Chile 102-104, 119, 211, 212
    Atacama Desert 120
    Coquimbo 103

China 7, 12, 13, 104, 107, 110, 155, 185, 187, 221
    Inner Mongolia 104
    Shansi 13
    Yellow River 110, 112, 185, 188

Climate change 4, 96, 97

Climax vegetation, change in 31

Cobar 121

Colorado 203, 204

Colorado River 201, 204, 205

Columbia (River) 112

Communities 125-127

Coquimbo 103

Cropping hazards 106, 107

Death Valley 5

Deforestation 9, 98, 180

Desert creep 7, 30, 37

Desert encroachment (see desert creep)

Desertification
    agricultural production 21, 22
    by continents 175

by countries
    Africa   183, 184
    Asia   189, 190
    Australia   198
    North America   207
    South America   214
    Spain   219
  cause   96-135
  combatting   23, 24, 137-143, 146, 164, 165, 166
    economic returns   23, 24
  control   (see combatting desertification)   160-164
  criteria   15, 16
  effects   96-135
  extent   15, 19
  factor   15
  impact   19
  indicators (see Indicators, desertification)
  irreversible   101, 164
  land area   15, 18
  maps
    Africa   176
    Asia   186
    Australia   193
    North America   200

INDEX 231

    South America   210

    Spain   216

  population affected   19, 20, 21

  processes   7, 28-57, 91

Desert spread (see desert creep)

Drought   2, 3, 4, 7, 8, 14, 72, 73, 100, 130, 155, 174, 181, 206, 211, 217

  Sahel   174, 178, 179

Dryland cropping (see dryland farming)

Dryland farming   15, 22, 23, 36, 41-46, 63-65, 105-110, 127, 154, 155, 157, 165, 173, 191, 212, 217, 220, 221

Dune encroachment   32, 33, 102

Dust Bowl   203

Ebro River   215, 217

Edaphic aridity   30, 31

Ekrafane Ranch   8

Egypt   117-119, 130, 179, 180, 182, 221

   Aswan   117

   Egyptians   10

   Nile (River)   112, 117, 177, 179-181

Egyptians   10

Erosion   2, 4, 10, 11, 13, 15, 28, 31-33, 40-44, 51, 52, 54, 63, 64, 65, 74, 90, 102, 107, 108, 128, 130, 144,

177, 185, 192, 194, 196, 199, 201, 202, 206, 215, 217, 218

    accelerated erosion   31, 32, 37, 51, 54

    geologic erosion   31, 32

    water erosion   7, 9, 13, 31-33, 42-44, 90, 91, 98, 105, 107, 109, 130, 136, 154-156, 182, 185, 188, 191, 202, 204, 206, 208, 209, 213, 215, 217

    wind erosion   2, 7, 31-33, 42-45, 50, 89, 91, 107, 109, 130, 154, 156-158, 182, 187, 196, 202-204, 206, 208, 209, 212, 213

Ethiopia   182, 221

Euphrates   11, 113, 187

Europe   155

Fertile Crescent   187

Fire   4, 40, 149, 150, 161, 162

Firewood   39, 52, 126, 148, 180, 212

Fuelwood (see Firewood)

Gascoyne Basin   101, 194

Gezira   181

Gila River   201

Goats   147, 148, 161, 162

Gran Chaco   212, 213

Grazing land   15, 23, 36-41, 59-63, 97-106, 127, 138,
    140, 144-148, 165, 166, 173, 177, 179, 185, 199,
    201-206, 212, 215, 221
Great Plains   2, 157, 203, 204, 208

Herida   217

Imperial Valley   130, 204
India   130, 185, 191
    Rajasthan   130
Indicators, desertification   58-95
    biological indicators   74-82
    critical indicators   68, 69
    direct indicators   72, 73
    indirect indicators   72, 73, 81, 84
    interpretation (indicators)   88, 89, 90
    measurement (indicators)   88, 89, 90
    physical indicators   74-82
    socioeconomic indicators   82, 83, 90
Indus (River)   112, 116, 185, 191
Inner Mongolia   104
Iran   98, 187, 221
Iraq   11, 12, 113, 116, 191, 221
    Euphrates   11, 113, 187

Mesopotamia   8, 11, 12, 187

  Mussayeb   113, 114

  Tigris   11, 113, 187

Irrigation (see Irrigation agriculture)

Irrigation agriculture   36, 46-50, 54, 110

Irrigated land   15, 23, 65, 127, 136, 158, 165, 166, 173, 177, 185, 192, 194, 195, 197, 213, 215, 218, 220, 221

Kara Kum Canal   112, 130

Kara Kum Desert   105

Kopet Dagh Mountains   105

Kenya   182, 221

Land deterioration, annual   22, 23

Land pollution   28, 29

Lebanon   10, 187

Libya   9

Loess   12, 13, 109, 110, 188

Loessial plateau   8, 13, 185, 188

Los Angeles   124

Maghreb   182

Mali   3

Management practices   143-159

Manitoba   205

Mauritania   3

Mediterranean   7, 8, 9, 11, 185

Mesopotamia   8, 11, 12, 187

Mexicali Valley   130, 204

Mexico   129, 130, 199, 201, 204, 206, 208, 209, 221

    Baja California   201

    Chihuahua   202

    Sonora   201

    Sonoran Desert   201

Middle East   98, 107, 129, 185, 187, 191

    Fertile Crescent   187

Mining   36, 50, 51, 103, 119-122, 127, 151, 152, 166, 177

Mojave Desert   201

Mona   116, 117

Monitoring   68, 72, 81, 82, 84, 88, 90, 93

Morocco   221

Murray Basin   197

Murray Mallee   196

Murray River   194, 195

Mussayeb   113, 114

Navajo Indian Reservation   202

Negev Desert   98

New Mexico   202, 204

New South Wales   194-196

Niger   3, 99, 100, 178

   Ekrafane Ranch   8

Nile (River)   112, 117, 177, 179-181

North Africa   9, 129, 177, 180, 182

   Maghreb   182

North America   14, 155, 180, 199-209

North Dakota   205

Northern Territory   194

Off-road vehicles   51, 123-125, 153

Ord River   194

Overgrazing   3, 8, 12, 15, 28, 36-39, 96, 97, 99, 101, 102, 105, 129, 174, 179, 185, 187, 192, 194, 199, 201-203, 209, 212, 213, 217

Nutrient availability (see Soil fertility)

Pacific Northwest   203, 204

Pakistan   116, 117, 130, 185, 191, 221

   Indus (River)   112, 116, 185, 191

   Mona   116, 117

Palouse   204

Pecos River   112

Peru   209, 211, 212, 213

Phoenicians   9, 10

Plan of Action to Combat Desertification   3

Plant nutrients   (see Soil fertility)

Quadalqivir Valley   215

Rainfed croplands (see Dryland farming)

Rajasthan   130

Range condition   31, 59-62, 67

   classes   61

   trends   63, 67

Rangelands (see Grazing lands)

Recreation   36, 51, 122-125, 127, 152-154, 166

Reforestation   10, 40, 149

Remote sensing   84, 85

Rio Salado   112, 209

Rocky Mountains   32

Roman   9

Sahara   3, 4, 7, 14, 99, 129, 177, 181, 182, 220

Sahel   174, 178, 179

   drought   2, 3, 8, 14

Saline seeps   44, 45, 65, 109, 180, 192, 194, 195, 197, 199, 205, 208

Salinization (salinity)   7, 12, 15, 28, 33, 34, 46-49, 54, 65, 73, 74, 91, 110-119, 128, 130, 155, 158, 159, 177, 179-182, 185, 187, 191, 192, 194-197, 199, 201, 205, 208, 211, 213, 214, 215, 217, 220, 221

Sand blasting   44

Sand dunes   5, 23, 29, 32, 43, 105, 150, 202, 209, 215

Sand dune stabilization   150, 151, 191

San Joaquin Valley   204

Santiago del Estero   213

Saskatchewan   205

Satellite imagery   82, 85, 88, 94, 96

Saudi Arabia   187, 221

Senegal   3

Shansi   13

Sightseeing   125

Siltation   12

Sodium, adsorbed   49, 66, 160

Soil compaction   7, 15, 28, 34, 35, 41, 42, 45, 46, 49, 50, 54, 65, 91, 154, 158, 160, 205

Soil Conservation Service, U.S.   2, 208

Soil Crusting   41, 42, 45, 49, 50, 54, 65, 154, 158, 205

Soil degradation   91, 92, 93, 108

Soil fertility   43, 154, 158, 165, 177, 182, 187, 213, 218

Somalia   221

Sonora   201

Sonoran Desert   201

South America   14, 129, 155, 209-215

South Australia   194, 195

South Dakota   204, 205

Soviet Union (U.S.S.R.)   12, 104, 112, 130, 155, 157, 185

   Kara Kum Canal   112, 130

   Kara Kum Desert   105

   Kopet Dagh Mountains   105

   Syr Darya   112

   Turkmenistan (see Turkmen Republic)   112

   Turkmen Republic   104, 112

Spain   215-220

   Ebro River   215, 217

   Herida   217

   Quadalqivir Valley   215

   Zaragosa   217

Sudan   145, 181, 221

   Gezira   181

Syr Darya   112

Technology, role of   141, 142

Texas   112, 202

Tigris   11, 113, 187

Timber cutting (see Tree cutting)

Tourism   36, 51

Tree cutting   4, 10, 12

Tsetse fly   74, 129, 137, 138

Tunisia   9, 178-182

Turkmenistan (see Turkmen Republic)   112

Turkmen Republic   104, 112

Uganda   178

UNESCO major project   1, 2

United Nations Conference on Desertification   3, 5

United States   12, 32, 100-102, 120, 121, 129, 130, 136, 155-157, 199, 204, 205, 208, 209

   Arizona   201, 202

   Arizona Territory   202

     Navajo Indian Reservation   202

   California   124, 201, 204

     Death Valley   5

     Imperial Valley   130, 204

     Los Angeles   124

     Mojave Desert   201

   Colorado   203, 204

    Colorado River   201, 204, 205

    Columbia (River)   112

    Gila River   201

    New Mexico   202, 204

    North Dakota   205

    Pacific Northwest   203, 204

    Palouse   204

    Pecos River   112

    Rocky Mountains   32

    San Joaquin Valley   204

    South Dakota   204, 205

    Texas   112, 202

    Utah   204

    Wyoming   204

Upper Volta   3

Urban development   36, 53, 54

Utah   204

Vegetation degradation   (see Vegetative cover, degradation of)   59, 91, 93

Vegetative cover, degradation of   7, 14, 28, 29-31, 59, 91, 93

Vegetation destruction   3, 54

Victoria   195

Victoria River   194

Watering points   37, 52, 100, 147
Waterlogging   7, 15, 28, 33, 34, 47, 48, 54, 65, 73, 74, 110-119, 128, 130, 143, 158, 159, 177, 179-182, 185, 187, 191, 192, 195, 197, 199, 201, 204, 208, 209, 211, 213, 214, 215, 217, 220
West Africa   182
Western Australia   197
Woodcutting   8, 15, 39, 40, 52, 53, 96, 103, 126, 127, 148, 149, 161, 162, 180, 187, 209, 211, 212, 215
Wyoming   204

Yellow River   110, 112, 185, 188

Zaragosa   217